Organic Chemistry Concepts:
An EFL Approach

Organic Chemistry Concepts: An EFL Approach

Gregory Roos
Cathryn Roos

ELSEVIER

AMSTERDAM • BOSTON • HEIDELBERG • LONDON
NEW YORK • OXFORD • PARIS • SAN DIEGO
SAN FRANCISCO • SINGAPORE • SYDNEY • TOKYO

Academic Press is an imprint of Elsevier

Academic Press is an imprint of Elsevier
32 Jamestown Road, London NW1 7BY, UK
525 B Street, Suite 1800, San Diego, CA 92101-4495, USA
225 Wyman Street, Waltham, MA 02451, USA
The Boulevard, Langford Lane, Kidlington, Oxford OX5 1GB, UK

British Library Cataloguing-in-Publication Data
A catalogue record for this book is available from the British Library

Library of Congress Cataloguing-in-Publication Data
A catalog record for this book is available from the Library of Congress

ISBN: 978-0-12-801699-2

For information on all Academic Press publications
visit our web site at http://store.elsevier.com/

Printed and bound in USA

Working together
to grow libraries in
developing countries

ELSEVIER   Book Aid
            International

www.elsevier.com • www.bookaid.org

Contents

Preface

1. AIM

The purpose of this book is to show the main **concepts** of organic chemistry in a **simple, language-accessible** format. It is aimed at non-major students of chemistry who use **English as a foreign language (EFL)**.

Students often see organic chemistry as very different from, and much harder than other branches of chemistry. "Organic chemistry is a foreign language," they often say. "Organic chemistry is just memorizing."

This textbook addresses these issues by looking at the concepts needed to understand the many experimental facts. Unlike many textbooks which are written for specific degree programs such as Life Sciences, Medicine and Environmental Science, this textbook does not try to go from methane to DNA by listing tables of functional groups and lists of unrelated physical and chemical properties. Instead, this textbook starts with the core concepts and uses the specific molecules as examples to develop the concepts. This approach gives students a better understanding of the concepts that control the behavior of organic compounds. Later in their programs, students will find that this has given them a more solid grounding in the material.

2. CONTENT FEATURES

The key material in this textbook is delivered in an outline form for the student to expand, either during or after the course. Once they have the concepts and language tools of organic chemistry, they can work with relatively complex molecules. The topics are selected to address areas that usually cause problems for students.

The number of functional classes is purposely limited. The chapters and sections are ordered so that they build a **broad concept base** at this introductory level. A study of some natural product types is included to give students some complex molecules on which to use the concepts they have learned.

Each chapter in this textbook ends with a collection of **self-learning programs** interspersed with general questions. These frame-by-frame exercises are designed to let students develop their skills, as well as check their progress, with new concepts as they meet them.

3. LANGUAGE ACCESSIBILITY

Readability is another specific feature of this textbook. By keeping the language of this textbook as simple as possible, the cognitive load of reading and understanding in a foreign language is minimized, freeing up the students to better focus on the content. Grammar and vocabulary are kept as simple as possible. For example, virtually all verbs are in the present simple tense, and words like "because" are used consistently instead of variations such as "since," "due to," or "as a result of," By favoring repetition over variation, the non-native reader of English can more easily focus on and absorb the subject matter. Standard, straightforward sentence construction has been used, with linking words and phrases prominently placed to help guide the reader. Language analysis tools[1-3] show that the text is at a grade 9 reading level and has a reading ease score of 50–60. More than 99% of the nonsubject-specific technical words used in this book are drawn from the 2000 most common English words and the 570 most common academic words. All technical words related to organic chemistry are defined, and many are highlighted and collected in an easy-reference glossary.

1. http://www.editcentral.com/gwt1/EditCentral.html
2. http://www.online-utility.org/english/readability_test_and_improve.jsp
3. http://www.lextutor.ca/vp/

How to Use This Book

Bolded words are defined in the text. The **definitions** are collected in a glossary at the end of the book. When you see the word used again, you can refer to the glossary easily if you need to.

As you read about the concepts, you will see some examples that help you understand each concept better. However, chemical reactions are limited to the ones that show the underlying principle. Focus on the **type of reaction** and do not worry about the many variations which are possible. Some simple **reaction mechanisms** are described only when they are useful to the learning process.

Note that organic chemistry is a **three-dimensional science**. Therefore, you need to understand and practice the skill of drawing three-dimensional diagrams. Many of the diagrams in the book show you how to do this. For further help with this, refer to the appropriate **appendices** at the end of the book. If possible, you should try to use **molecular models**.

At the end of each chapter, there are **graded questions** for you to practice your skills. In addition, there are **self-learning programs** to help you understand the main concepts. The programs are made up of **question and answer frames**. Each one is designed to help you learn about a specific topic at your own speed. To get the full benefit from the self-learning programs, you should proceed as follows:

- Look only at the first frame (question frame) and try to write a full answer.
- Read the next frame to check your answer.
- The second frame may also ask the next question.
- Repeat the process as needed until you complete of the whole topic.
- **DO NOT MOVE ON UNTIL YOU UNDERSTAND THE CONCEPT COMPLETELY.**

Self-Learning Programs

CHAPTER 1
Organic Structures

1.1 WHAT IS ORGANIC CHEMISTRY?

Over the past 70 years, organic chemistry has become a very broad and complex subject. We see the results of this every day. There are new developments in food, pharmaceuticals, synthetic materials, and other petrochemical products. This progress is largely due to developments in modern instruments and theory. As a result, we can better understand the basic factors that control the behavior of organic compounds.

What is organic chemistry? New students usually answer: "The chemistry of carbon" or "The chemistry of life." Both of these are good answers, but why exactly can carbon play this special role?

1.2 WHAT MAKES CARBON SPECIAL?

Table 1.1 shows that carbon is one of the primary elements of life. Only carbon is able to form molecules with enough complexity to support life.

How important is each element of life? It does not only depend on quantity. But it does depend on the role it plays. For example, Table 1.1 shows that the human body has only a small amount of iron. However, iron is necessary for the hemoglobin to carry oxygen in the blood. Iodine is needed for the thyroid to work properly. Cobalt is part of vitamin B_{12}. Zinc, copper, and manganese are present in various enzymes. In each of these examples, there are many carbon, hydrogen, oxygen, and nitrogen atoms for each metal atom. However, without the **trace element** metals, it is impossible for these compounds to carry out their biological functions.

There are more than 30 million carbon-based compounds that are known so far. This number continues to grow every year. Why are carbon and its compounds such an important part of chemistry?

Table 1.1	Composition of the Human Body		
Element	**% by weight**	**Element**	**% by weight**
Oxygen	65	Sulfur	0.2
Carbon	18	Sodium	0.1
Hydrogen	10	Chlorine	0.1
Nitrogen	3	Magnesium	0.05
Calcium	2	Br, I	Traces
Phosphorus	1.1	Fe, Mg, Zn	Traces
Potassium	0.3	Cu, Co	Traces

Organic Chemistry Concepts: An EFL Approach. http://dx.doi.org/10.1016/B978-0-12-801699-2.00001-8

■ The **tetravalent** nature of the bonding of carbon. This means that carbon needs four bonds to complete an **octet** of electrons, in other words to fill its **valence** outer shell.

■ The ability to form strong **single covalent** bonds where the bonded atoms share an electron pair. Carbon atoms can bond in this way to an almost unlimited number of other carbon atoms. For **acyclic** compounds there are no rings. This gives either **straight chains** which have no branch points, or **branched chains** which do have branch points. In **cyclic** compounds there can be different sized rings.

■ The ability to form double or triple **multiple bonds**, where more than one electron pair is shared with another carbon atom.

■ The ability to bond covalently with many **heteroatoms**, other non-carbon atomic species such as H, O, N, S, P, and halogens. These bonds are either single or multiple.

1.3 MOLECULES, FORMULAE, AND STRUCTURES

Carbon can be part of different bonding arrangements in the group of bonded atoms that form a **molecule**. Because a molecular **formula** only gives the type and number of atoms in a molecule, it does not tell anything about the **structure** of the molecule. The structure gives information of how the atoms are joined together. For example, 366,319 structures with a molecular formula of $C_{20}H_{42}$ are possible. To simplify this problem, it is necessary to classify and subclassify organic substances.

The best place to start is with **hydrocarbons**, which are compounds that contain only carbon and hydrogen. Figure 1.1 shows how related structures and properties are used to classify hydrocarbons. As a first stage, hydrocarbons can be separated into **aromatic** or **aliphatic** types. All aromatic compounds have special bonding arrangement within a ring. You will see details of this aromatic bonding in later chapters. The word "aliphatic" then refers to all non-aromatic examples. Aliphatic hydrocarbons can be either **saturated** or **unsaturated**. Saturated compounds have no multiple bonds. Unsaturated compounds have at least one multiple bond.

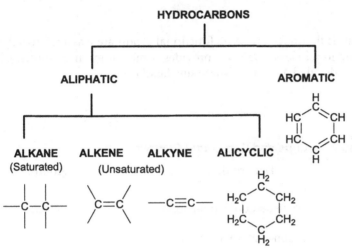

FIGURE 1.1
Primary classifications of hydrocarbon compounds.

Organic chemistry uses a number of special words that are not used in other branches of chemistry. Do not worry about this. These words will become familiar as you use them again and again. However, it is important to note that these words have specific meanings, and you must use them correctly.

Other common definitions that help with classifications are shown in Figure 1.2. These are:

- **acyclic** – structures that do not have a ring in them;
- **carbocyclic** – a ring that is made of only carbon atoms;
- **heterocyclic** – a ring that has at least one non-carbon atom in it.

Functional groups are an important way to classify organic compounds. Functional groups are fixed arrangements of atoms within a compound. These groups are mainly responsible for the physical and chemical properties of a compound. They are formed when carbon–hydrogen bonds in saturated hydrocarbons are replaced to give either multiple bonds or bonds to heteroatoms.

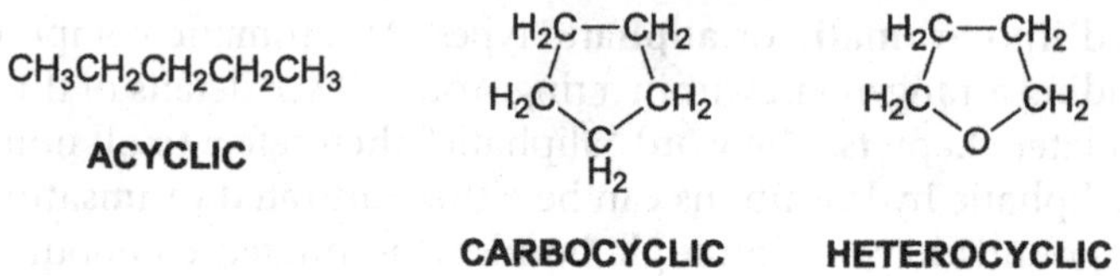

$$CH_3CH_2CH_2CH_2CH_3$$

ACYCLIC

CARBOCYCLIC   **HETEROCYCLIC**

FIGURE 1.2
Acyclic and cyclic classifications.

Compounds that have the same functional group are classified together in the same **functional class**. Table 1.2 provides some common examples. Chapter 2 provides a detailed account of these subclassifications.

Table 1.2	Common Functional Groups and Compound Classes	
Functional group	**Description**	**Compound class**
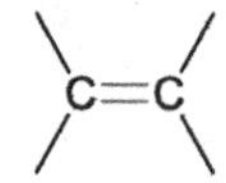 C=C	Carbon–carbon double bond	Alkene
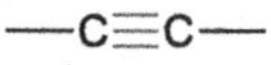 —C≡C—	Carbon–carbon triple bond	Alkyne
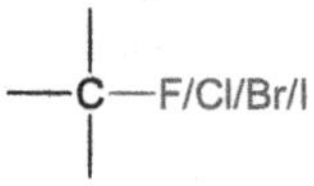 —C—F/Cl/Br/I	Halogen atom	Alkyl halide
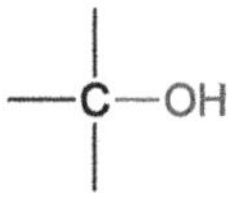 —C—OH	Hydroxyl group	Alcohol
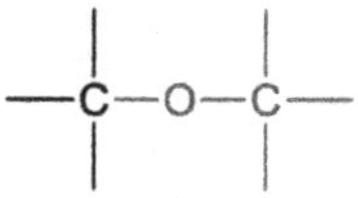 —C—O—C—	Alkoxy group	Ether
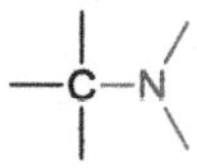 —C—N	Amino group	Amine
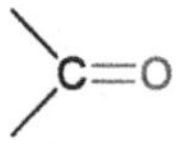 C=O	Carbonyl group	Aldehyde/ketone
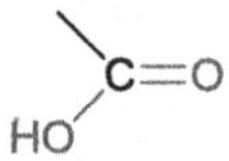 C=O, HO	Carboxyl group	Carboxylic acid
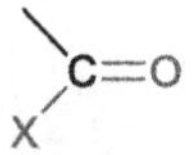 C=O, X X = various	Acyl group	Carboxylic acid derivatives

1.4 BONDS AND SHAPE: THE HYBRIDIZATION MODEL

To understand organic chemistry, we must understand bonding and shape, especially that of carbon. At this level of study, we can use the simple **hybridization** model to explain single and multiple bonding, as well as molecular shape. Hybridization is the mixing of atomic orbitals to give new **hybrid atomic orbitals** which have new shape and directional properties. These hybrid atomic orbitals then combine with other atomic orbitals to form the bonds in molecules.

Table 1.3 Hybridization States of Carbon			
	sp	*sp²*	*sp³*
Number of orbitals	2	3	4
Interorbital angle	180°	120°	109.5°
Orbital arrangement	Linear	Trigonal	Tetrahedral
Remaining *p* orbitals	2	1	0
Bonds formed	$2\sigma, 2\pi$	$3\sigma, 1\pi$	4σ
% s character	50	33⅓	25
% p character	50	66⅔	75
Carbon electronegativity	3.29	2.75	2.48
C–C bond length (pm)	121	133	154
Average C–C bond energy (kJ/mol)	837	620	347

Carbon has one $2s$ and three $2p$ orbitals for use in hybridization. Table 1.3 shows that the combination of the $2s$ orbital with three, two, or one $2p$ orbital leads to $4sp^3$, $3sp^2$, and $2sp$ hybrid atomic orbitals. Figures 1.3–1.5 show that all three of these results give the tetravalency that carbon needs by allowing for single or multiple bonds to be present.

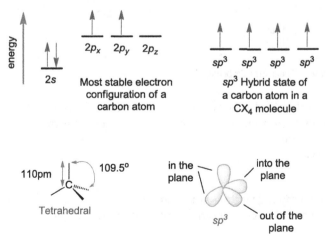

FIGURE 1.3
sp^3-hybridized carbon (tetrahedral, four single σ bonds).

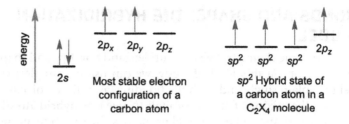

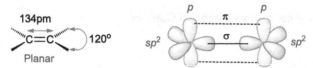

FIGURE 1.4
sp^2-hybridized carbon (trigonal, $3\sigma + 1\pi$ bonds).

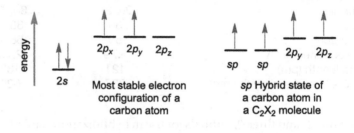

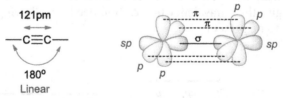

FIGURE 1.5
sp-hybridized carbon (linear, $2\sigma + 2\pi$ bonds).

Because the s orbital is lower in energy and closer to the nucleus than p orbital, hybrid orbitals with a greater percentage of s character form shorter, stronger bonds. Also, as the s orbital content increases, both the bond length and bond energy decrease.

> Hybridization must give the same number of new hybrid atomic orbitals as the number of original atomic orbitals that are combined.

The **sigma** (σ) and **pi** (π) types of covalent bonds come from the relative direction of the axes of the overlapping bonding atomic orbitals. A σ bond has direct overlap along the orbital axis. This gives a bonding orbital that is cylindrically symmetrical. A π bond results from the less efficient sideways overlap of orbitals that are in the same plane.

We can estimate the strength of the π bond as about 273 kJ/mol by using the bond energies of the C–C and C=C as given in Table 1.2. Therefore it is much weaker than σ bond (347 kJ/mol). This fact is important because it explains the higher reactivity of multiple bonds.

To find the hybrid state of any carbon atom, simply count the number of different atoms bonded directly to it. An sp^3 carbon bonds to four other atoms with single σ bonds. An sp^2 carbon bonds to three other atoms with two single and one double bond. An sp carbon bonds to only two other atoms with one single and one triple bond or two double bonds.

1.5 POLAR BONDS AND ELECTRONEGATIVITY

The polarity of a chemical bond shows how the bonding electrons are shared between the bonded atoms. Figure 1.6 shows the range from the extremes of ionic, between anions and cations, and perfect covalent, in which identical atoms or groups share the bonding electrons equally. All situations between these are examples of polar covalent bonding.

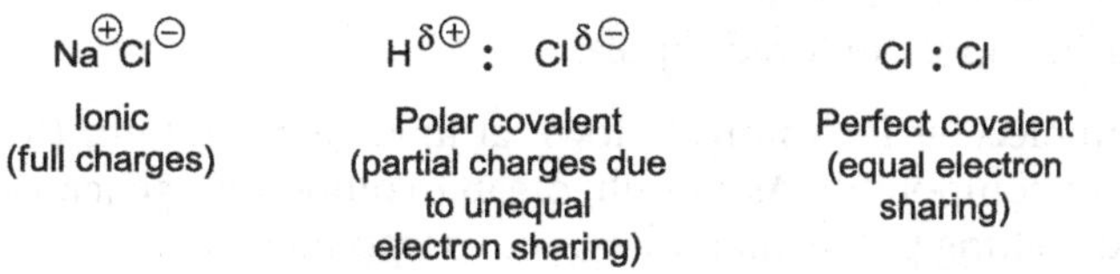

$$Na^{\oplus}Cl^{\ominus} \qquad H^{\delta\oplus} : Cl^{\delta\ominus} \qquad Cl : Cl$$

Ionic (full charges) Polar covalent (partial charges due to unequal electron sharing) Perfect covalent (equal electron sharing)

FIGURE 1.6
The bonding range from ionic to covalent. The symbol δ is often used to show a partial/small amount of charge.

In **polar bonds** one nucleus attracts the bonding electrons more than the other. **Electronegativity** measures the attraction which a bonded atom has for the bonding electrons. As the electronegativity difference between the bonded atoms increases, the polar character of the bond between them increases. Further details and values are listed in Appendix 1.

In organic chemistry, we talk about the polarity of a bond in terms of the **inductive effect (I)**. This shows the ability and direction with which an atom or group of atoms polarizes a covalent bond by donating or withdrawing electron density.

The most interesting bonding centers are usually carbon. As Figure 1.7 shows, it is usual to indicate an inductive effect relative to the almost non-polar C–H bond. The effect of other atoms or groups is then expressed as ±I.

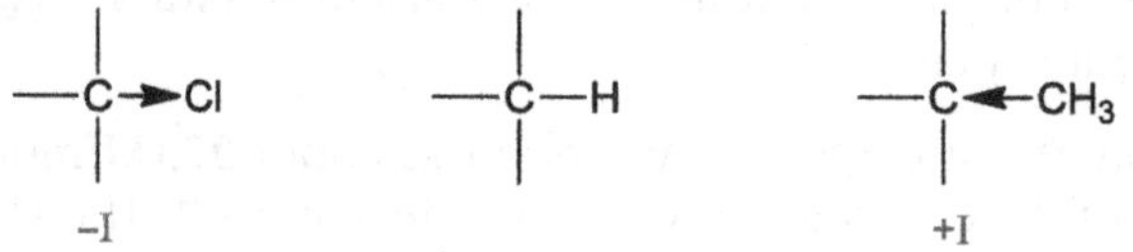

$$-\overset{|}{\underset{|}{C}}\!\!\rightarrow\! Cl \qquad -\overset{|}{\underset{|}{C}}\!\!-\! H \qquad -\overset{|}{\underset{|}{C}}\!\!\leftarrow\! CH_3$$

$$-I \qquad\qquad\qquad\qquad\qquad +I$$

FIGURE 1.7
Negative and positive inductive effects of carbon.

Note that an inductive effect refers to σ-bonded electrons only. The σ-bonded electrons are **localized**. This means that they are found mostly between the bonded nuclei. Because of this, an inductive effect is only felt over very short distances, and is almost gone after one bond. Later chapters use the inductive effect in discussions of molecular properties and reactivity.

1.6 FORCES BETWEEN MOLECULES

In ionic compounds, electrostatic attraction causes the ions to form large three-dimensional arrangements called crystals. For organic compounds, in which the bonding is mostly covalent, the unit is usually an uncharged single molecule. The relatively weak attractive **intermolecular** interactions, the van der Waals forces, between these molecules are of three types:

- dipole/dipole (includes hydrogen bonding)
- dipole/induced-dipole
- induced-dipole/induced-dipole.

These intermolecular forces break down at lower temperatures (lower energy) than for ionic compounds. As a result, organic compounds generally have lower boiling and melting points than inorganic compounds.

The strength of the intermolecular interactions depends on the polarization of various parts of the organic molecule. One cause of polarization is the inductive effects that come from the presence of electronegative heteroatoms. This polarization leads to dipole/dipole interactions. Also, a dipole can affect the electron field in a part of any nearby molecule. This can cause an induced-dipole to form and lead to dipole/induced-dipole interactions.

Even non-polar molecules can have temporary distortions in their electron fields. These short-lived induced-dipoles can cause distortions in a part of other nearby molecules. As shown in Figure 1.8, this can lead to induced-dipole/induced-dipole interactions. Extended induced-dipole/induced-dipole interactions over many molecules can add up to give significant intermolecular attraction.

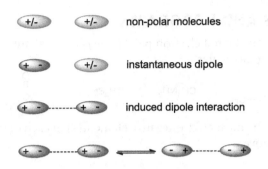

FIGURE 1.8
Induced-dipole/induced-dipole intermolecular forces.

Generally, as molecular size increases, so does the total van der Waals interaction. The efficiency of this attraction can also depend on molecular shape, and how well the molecules can fit together. Therefore, as chain-branching increases, the efficiency of the van der Waals interaction between molecules decreases as shown in Figure 1.9.

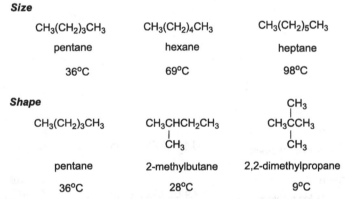

FIGURE 1.9
The dependence of intermolecular forces on size and shape.

The polarity and type of intermolecular interactions of organic molecules can also explain their solubility properties. Organic compounds generally have low solubility in polar solvents like water. This is because they are either non-polar or only moderate polar. This means they have little attractive interaction with the solvent molecules. In contrast, ionic compounds can ionize and polar solvent molecules can interact strongly with the ions. This interaction, called **solvation**, makes the ion more stable and helps with solubility.

QUESTIONS AND PROGRAMS

Q 1.1. Draw the unshared electron pairs (*lone pairs*) that are missing from the following molecules.

$$H_3C\diagdown O{-}CH_3 \qquad CH_3NH_2 \qquad CH_3Br \qquad H_3C{-}\overset{\overset{O}{\|}}{C}{-}OCH_3$$

Q 1.2. Identify the most electronegative element(s) in each of the molecules in Q 1.1 above.

PROGRAM 1 Percentage Ionic Character of Covalent Bonds

A Covalent bond polarity can be given as a percentage of ionic character. This is calculated from the electronegativities E_a and E_b of the more and less electronegative atoms in the bond.

$$\% \text{ Ionic character} = \frac{E_a - E_b}{E_a} \times 100$$

Use the values in Appendix 1 to calculate the ionic character of the covalent single bonds in C–O, C–H, and O–H. Show the partial charges for each bond.

B Simple calculation gives:

C—O

$$\% \text{ Ionic character} = \frac{3.44 - 2.55}{3.44} \times 100 = 26\%$$

$$\overset{\delta\oplus\ \delta\ominus}{C{-}O}$$

C—H

$$\% \text{ Ionic character} = \frac{2.55 - 2.21}{2.55} \times 100 = 13\%$$

$$\overset{\delta\ominus\ \delta\oplus}{C{-}H}$$

O—H

$$\% \text{ Ionic character} = \frac{3.44 - 2.21}{3.44} \times 100 = 36\%$$

$$\overset{\delta\ominus\ \delta\oplus}{O{-}H}$$

Q 1.3. In each of the following sets, arrange the covalent bonds in an order of increasing partial ionic character (i.e., *increasing polarity*).

(a) C–H, O–H, N–H (b) C–H, B–H, O–H (c) C–S, C–O, C–N

(d) C–Cl, C–H, C–I (e) C–N, C–F, B–H (f) C–Li, C–B, C–Mg

Q 1.4. Study the following molecules and name the functional class for each.

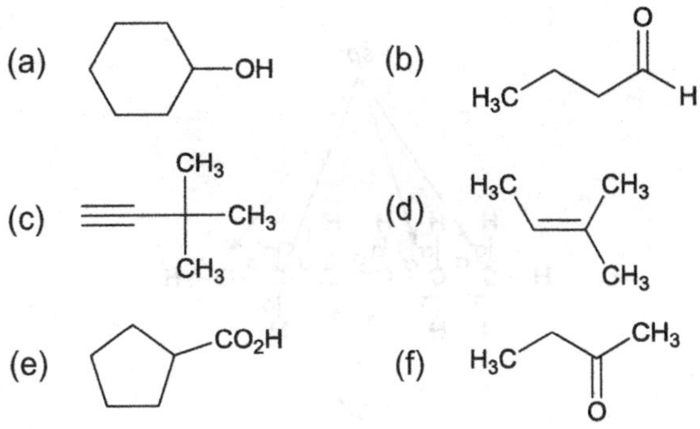

(a) [cyclohexane ring]—OH   (b) [structure: H₃C—CH₂—CH₂—CHO] H_3C ... H

(c) [HC≡C—C(CH₃)(CH₃)—CH₃ structure with CH₃ groups]   (d) [structure with H_3C, CH_3, CH_3]

(e) [cyclopentane ring]—CO₂H   (f) [structure H_3C ... CH_3, O]

Q 1.5. Show the hybridization state of the non-hydrogen atoms in the following molecules.

[structures: CH_3CH_2—C(=O)—OH, H_2C=CH—C(=O)—CH_3, $HC\equiv C$—CH_2—C=N]

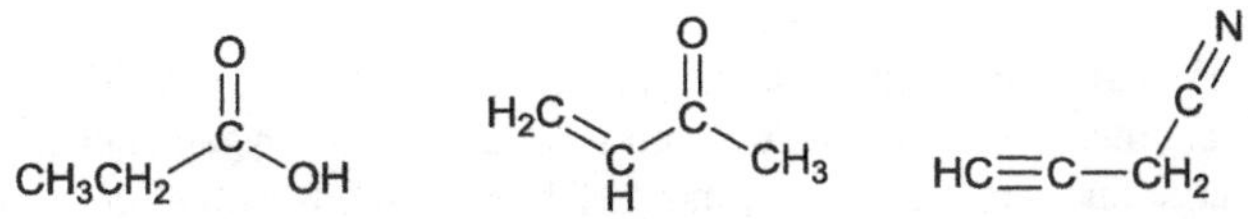

Q 1.6. Draw orbital diagrams to show the bonding in the following molecules.

CH_3—C(=O)—OH H_2C=CH—CH_2Br CH_3—$C\equiv N$

PROGRAM 2 Molecular Structural Features

A   Study the following molecular structure and write down as much structural information as you can. (*Hint*: functional groups, bonding, classifications, shape, etc.).

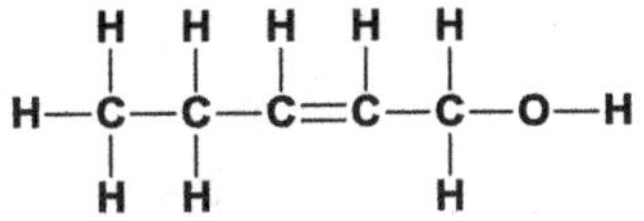

[structure: H—C—C—C=C—C—O—H with H atoms]

B   At first, you should at least have identified the functional groups of the alkene C=C and the alcohol C–OH (hydroxyl function containing an oxygen heteroatom) on an acyclic skeleton.

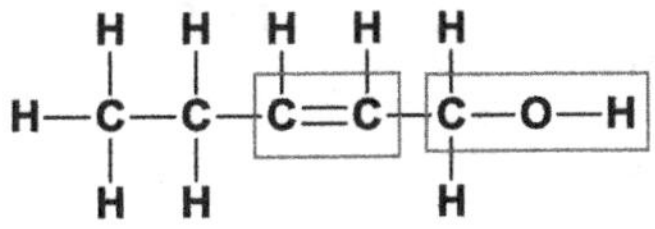

[structure: H—C—C—|C=C|—|C—O—H| with H atoms, boxes highlighting groups]

Now dig deeper.

C A closer look shows the hybrid state of the C and O atoms. This allows the bonding to be classified as σ (15 of these) or π (1 of these) bonds.

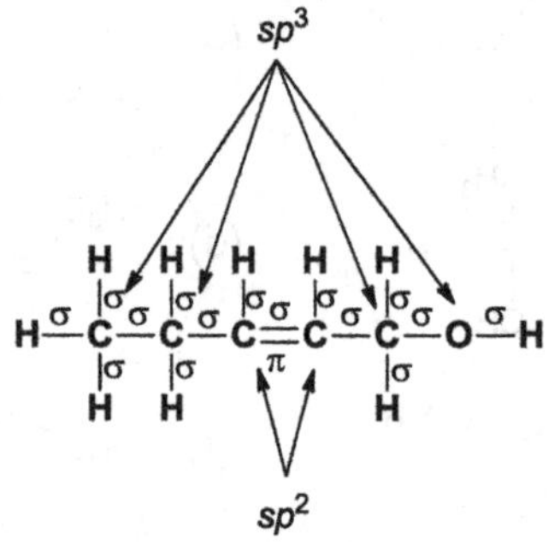

sp^3

$$H-\overset{\overset{\displaystyle H}{|}}{\underset{\underset{\displaystyle H}{|}}{C}}-\overset{\overset{\displaystyle H}{|}}{\underset{\underset{\displaystyle H}{|}}{C}}-\overset{\overset{\displaystyle H}{|}}{C}=\overset{\overset{\displaystyle H}{|}}{C}-\overset{\overset{\displaystyle H}{|}}{\underset{\underset{\displaystyle H}{|}}{C}}-O-H$$

sp^2

🐌 Do not stop here. Look even harder.

D Some additional things include: the oxygen lone pairs; the tetrahedral (sp^3) and trigonal (sp^2) shapes; the polar bonds to the electronegative oxygen (inductive effect); the four **coplanar** carbons, because of the flat shape of the C=C carbon atoms.

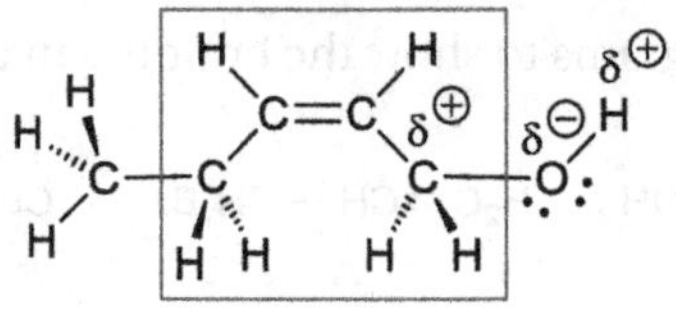

Q 1.7. Apply Program 2 mentioned above to the following molecular structures.

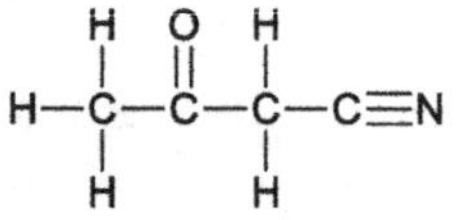

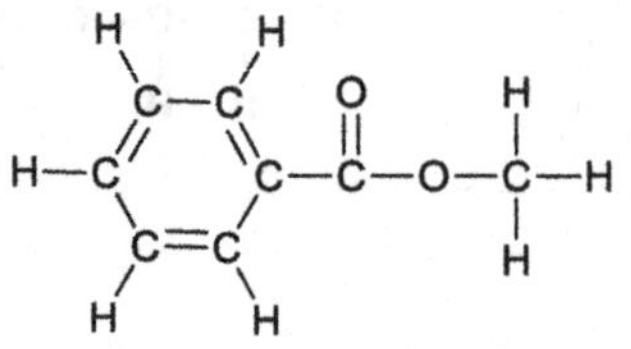

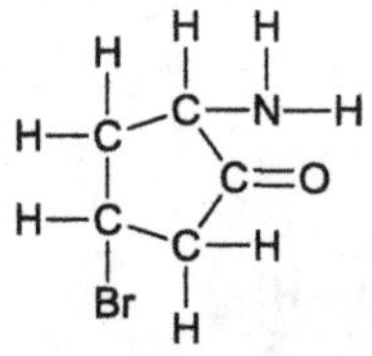

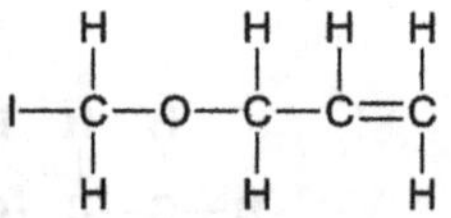

Q 1.8. Write down the molecular formulae for the molecules mentioned in Q 1.7.

PROGRAM 3 Intermolecular Forces

A   The forces of attraction between particles (atoms, ions, molecules) are electrostatic. However, these interactions are very different in their relative strength.

The strongest attraction is between ions. For example, the interaction between Na^+ and Cl^- is 787 kJ/mol. The attraction between permanent dipoles is next strongest at 8–42 kJ/mol. Finally, the weakest interaction of 0.1–8 kJ/mol is between induced dipoles.

The forces between the molecules of organic compounds are mostly of the last two types. This explains their relatively low melting and boiling points. The size of the temporary induced dipoles depends directly on molecular size.

🖎 Study the following set of unbranched hydrocarbons and try to arrange them in order of increasing boiling point.

$CH_3CH_2CH_2CH_3$	$CH_3(CH_2)_4CH_3$	$CH_3(CH_2)_8CH_3$
Butane	Hexane	Decane

B   All three molecules are unbranched hydrocarbons. Therefore, the attractive forces depend directly on molecular size, and so the order is unchanged.

$CH_3CH_2CH_2CH_3$	$CH_3(CH_2)_4CH_3$	$CH_3(CH_2)_8CH_3$
Butane Bp −0.5°C	Hexane Bp 69°C	Decane Bp 174°C

🖎 What is the order for the following structural isomers?

$$CH_3(CH_2)_6CH_3 \qquad CH_3\!-\!\overset{\overset{\displaystyle CH_3}{|}}{\underset{\underset{\displaystyle CH_3}{|}}{C}}CH_2\overset{\overset{\displaystyle CH_3}{|}}{C}HCH_3 \qquad CH_3\!-\!\underset{\underset{\displaystyle CH_3}{|}}{C}H(CH_2)_4CH_3$$

Octane	2,2,4-Trimethylpentane	2-Methylheptane

C   All three hydrocarbons have the same C_8H_{18} molecular formula. So size alone cannot determine the attractive forces. The molecular shape, which is given by the amount of branching, is important. This determines the effective surface area of the molecules. As branching increases, the effective surface area decreases, and the forces of attraction decrease. This shows the ability of the molecules to pack in well-ordered arrays. Therefore the order is:

$$CH_3\!-\!\overset{\overset{\displaystyle CH_3}{|}}{\underset{\underset{\displaystyle CH_3}{|}}{C}}CH_2\overset{\overset{\displaystyle CH_3}{|}}{C}HCH_3 \qquad CH_3\!-\!\underset{\underset{\displaystyle CH_3}{|}}{C}H(CH_2)_4CH_3 \qquad CH_3(CH_2)_6CH_3$$

2,2,4-Trimethylpentane	2-Methylheptane	Octane
Bp 99°C	Bp 118°C	Bp 126°C

D Now consider some compounds that have relatively strong permanent dipoles because of highly polarized bonds. Hydrogen bonding, at $\pm 20\,kJ/mol$, is the strongest of these forces. This occurs wherever a hydrogen atom is bonded to a very electronegative element, most commonly F, O, or N. This relatively strong interaction has a large effect on properties such as boiling point and solubility.

CH_3CH_2Cl	CH_3CH_2OH	CH_3OCH_3	$CH_3CH_2CH_3$
Chloroethane	Ethanol	Dimethyl ether	Propane
MW 64.5	MW 46	MW 46	MW 44

🐟 Arrange the above compounds in order of increasing boiling point.

E You should have identified the non-polar alkane as having the weakest attractive forces. The alcohol has the strongest attractive forces because of hydrogen bonding. The ether and the alkyl halide lie between these extremes based on their polar bonds and their relative molecular weights.

$CH_3CH_2CH_3$	CH_3OCH_3	CH_3CH_2Cl	CH_3CH_2OH
Propane	Dimethyl ether	Chloroethane	Ethanol
MW 44	MW 46	MW 64.5	MW 46
Bp $-42^{\circ}C$	Bp $-24^{\circ}C$	Bp $12^{\circ}C$	Bp $78^{\circ}C$

Hydrogen bonding with water molecules is the reason that small alcohols and polyhydroxy alcohols have good solubility in water.

CHAPTER 2

Functional Classes I, Structure and Naming

2.1 DRAWING AND NAMING MOLECULES

To understand the chemistry of organic molecules, we need to know the types of compounds that are possible. In this chapter we look at some details of the important functional classes introduced in Chapter 1. Each compound class is shown with structural diagrams (how to draw the compounds) and systematic naming of the compounds. This background knowledge will prepare you for the chemistry in later chapters.

2.2 SATURATED HYDROCARBONS

Hydrocarbon means that this class of compound has only carbon and hydrogen. In this broad grouping there are both:

- acyclic examples called alkanes;
- cyclic examples called cycloalkanes.

All saturated examples have only single σ-bonds between sp^3-hybridized carbon atoms and hydrogen atoms. This class gives the parent compounds from which all other functional types come from. They also serve as the parent compounds for systematic naming.

Hydrocarbons have low chemical reactivity. This is because they have no reactive functional group. They simply consist of chains of tetrahedral carbon atoms which are surrounded by hydrogen atoms. Table 2.1 gives a selection of hydrocarbons along with their physical properties of melting and boiling points. These low melting and boiling values show their overall non-polar character. Hydrocarbons can have "straight" chains (do not forget the shape caused by the tetrahedral carbon), branched chains, and cyclic variations.

For any of these subclasses, we can write a series of compounds that have the same basic structure, but differ from each other by a single extra –CH_2– methylene group. Any series of compounds like these is called a **homologous series** and its members are homologs of each other.

2.2.1 Structural Diagrams

The purpose of a structural diagram is to show details for the arrangement of atoms in a particular compound. As shown in Figure 2.1, there are a number

Organic Chemistry Concepts: An EFL Approach. http://dx.doi.org/10.1016/B978-0-12-801699-2.00002-X

Table 2.1   Parent Acyclic Alkanes and Cycloalkanes

IUPAC Name	Molecular Formula	Structural Formula	M.P. (°C)	B.P. (°C)
Methane	CH_4	CH_4	−182	−162
Ethane	C_2H_6	CH_3CH_3	−183	−89
Propane	C_3H_8	$CH_3CH_2CH_3$	−187	−42
Butane	C_4H_{10}	$CH_3(CH_2)_2CH_3$	−135	−0.5
Pentane	C_5H_{12}	$CH_3(CH_2)_3CH_3$	−130	36
Hexane	C_6H_{14}	$CH_3(CH_2)_4CH_3$	−94	69
Heptane	C_7H_{16}	$CH_3(CH_2)_5CH_3$	−91	98
Octane	C_8H_{18}	$CH_3(CH_2)_6CH_3$	−57	126
Nonane	C_9H_{20}	$CH_3(CH_2)_7CH_3$	−54	151
Decane	$C_{10}H_{22}$	$CH_3(CH_2)_8CH_3$	−30	174
Cyclopropane	C_3H_6	△	−127	−33
Cyclobutane	C_4H_8	□	−80	−13
Cyclopentane	C_5H_{10}	⬠	−194	49
Cyclohexane	C_6H_{12}	⬡	6.5	81

IUPAC, International Union of Pure and Applied Chemistry.

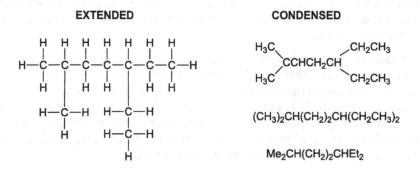

FIGURE 2.1
Structural diagrams.

of ways to do this. The choice of method depends on the specific structural feature(s) of interest.

For the beginner, the full Lewis-type structure (extended) is the safest choice. Because every bond and atom is shown, we can avoid mistakes with the tetravalent nature of carbon. After practice with examples that have different structural features and functional groups, it becomes easier to use the shorter forms, such as condensed and bond line types.

The condensed forms use groups of atoms and show almost no detail of individual bonds. These groups can show all atoms, for example CH_3- and $-CH_2-$. Alternatively, accepted short forms can be used, for example Me– for CH_3- and Et– for CH_3CH_2-. Often it is useful to use a combination of structural diagram forms. In these diagrams, only important features are shown in full detail.

You must take care to draw any bonds between the actual bonded atoms. This will avoid any mistakes with the valency (oxidation state) of the atoms involved. Note that only the bond line method shows the shape of the carbon framework. This is because every bend in the diagram represents a bonded group, for example $-CH_2-$. The ends of lines represent CH_3- groups.

It is also useful to be able to describe the degree of substitution at saturated sp^3 carbon centers. This is simply done by counting the number of hydrogen atoms bonded to the particular carbon. As Figure 2.2 shows, this gives rise to four types:

- primary, with 3 *H*s on carbon;
- secondary, with 2 *H*s on carbon;
- tertiary, with 1 *H* on carbon;
- quaternary, with no *H*s on carbon.

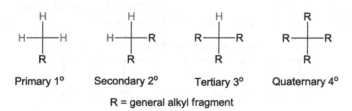

Primary 1°   Secondary 2°   Tertiary 3°   Quaternary 4°

R = general alkyl fragment

FIGURE 2.2
Classification of carbon centers.

It is also common to use the symbol –R to show general **alkyl groups**. A selection of these are detailed in Section 2.2.3 and are derived from alkanes by removing a hydrogen ligand.

In addition, as Figure 2.3 shows, there are different ways to show the **three-dimensional** (3-D) shape of tetrahedral sp^3 centers. A tetrahedral center has four **substituents**, or attached groups. The most common is to show two adjacent

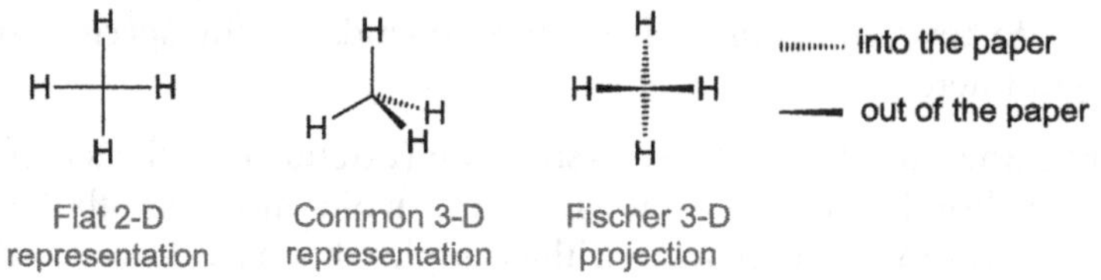

H
H———H
H

Flat 2-D
representation

H
H   H
H

Common 3-D
representation

H
H———H
H

Fischer 3-D
projection

··········· into the paper

———— out of the paper

FIGURE 2.3
Three-dimensional representations around atomic centers.

substituents in the plane of the paper with normal bond lines. The other two substituents are drawn going into the paper with a dashed wedged bond, or coming out of the paper with a solid wedged bond.

The Fischer projection is a less common alternative. By definition in these drawings, the vertical bonds go into the paper, and the horizontal bonds come out of the paper.

You do not always have to show the full **stereochemistry** (3-D shape) of a molecule. However, as you will see in Chapter 3, it is important not to forget that molecules have 3-D shapes.

2.2.2 Oxidation States for Carbon

This concept helps to create a link between the various classes of carbon compounds. The type and electronegativity of the atoms which are bonded to a carbon lets us assign nominal **oxidation numbers** to the various carbon atoms. These oxidation numbers indicate the relative gain or loss of electrons at the carbon in each compound type. This shows the relative equivalence of particular carbon oxidation states. From this, we can compare the oxidation levels of different functional groups.

The series of oxygen-containing functional classes in Figure 2.4 shows the principle. We can extend this process to other functional classes that involve other heteroatoms such as nitrogen, sulfur, and the halogens.

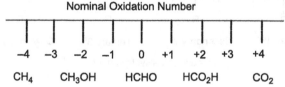

Nominal Oxidation Number

| −4 | −3 | −2 | −1 | 0 | +1 | +2 | +3 | +4 |

CH_4 CH_3OH HCHO HCO_2H CO_2

FIGURE 2.4
Nominal carbon oxidation numbers in functional classes.

Hydrogen is given the oxidation number of +1. Therefore, methane has carbon in its most reduced form of −4, which is its most stable, least reactive state. If a hydrogen atom is replaced with a bond to another carbon, the nominal oxidation number of the original carbon changes to −3. This is because we consider the carbons to have no effect on each other. The replacement of another hydrogen atom with a carbon, or the formation of a carbon–carbon double bond, then changes the oxidation number to −2, and so on.

Hydrocarbons (alkanes, alkenes, alkynes) can have carbons with nominal oxidation numbers ranging from –4 to 0. This depends on the number of other carbons attached. This follows the sequence from methane through 1°, 2°, 3°, and 4° carbon centers as was shown in Section 2.2.1. This helps us understand the different characteristics which they show in their reactions.

When we apply this process to common heteroatoms, they are all more electronegative than carbon and will count as –1 per bond. Therefore, the alcohol in Figure 2.4 has the functional group carbon with a –2 oxidation number. This comes from the +3 for the hydrogens bonded to the carbon and –1 for the single bond to oxygen. The aldehyde, with two bonds to oxygen, has the carbon with a 0 oxidation state. This is made up of +2 for the hydrogens and –2 for the two oxygen bonds. We can use the same process for carbon in its most oxidized form of +4 in CO_2.

This concept also helps us understand a number of other basic concepts. These include organic reactions in Chapter 5 and the acid/base properties of organic molecules in Chapter 6.

2.2.3 Systematic Naming for Alkanes

Chemical naming is needed for the accurate communication of structural information. The International Union of Pure and Applied Chemistry (**IUPAC**) is responsible for the system of naming chemical compounds. The IUPAC system provides the formal framework for naming. However, many common historical names are still used, and these are best learned through experience.

The full rules of IUPAC naming fill many hundreds of pages. It is not practical or necessary to cover all of this. Below are the general rules for substitutive naming of alkanes. This approach is based on replacing hydrogen with other groups.

- Identify the major functional group present. This gives the class name and name ending—in this case -ane for alkane and cycloalkane.
- Find the longest continuous carbon chain which has the functional group in it. This provides the parent name.
- Number the chain so that the functional group gets the lowest possible number. For saturated hydrocarbons the direction of the numbering depends on the position of any substituents.
- Identify all substituents and their numerical positions on the chain. For saturated hydrocarbons, the chain is numbered so that substituents have the lowest set of possible numbers.
- Note any possible stereochemical requirements. In this book, this only applies to cycloalkanes and alkenes in which the labels *cis/trans* and *E/Z* are used as needed.
- Put the above information together by listing the substituents and their chain positions, in alphabetical order, ahead of the parent class name. Numbers are separated by commas and words are separated from numbers by hyphens.

Table 2.2 Common Alkyl Groups		
Alkyl Group	**IUPAC Name**	**Alternative Contractions**
CH_3-	Methyl	Me–
CH_3CH_2-	Ethyl	Et–
$CH_3CH_2CH_2-$	Propyl	Pr–
$CH_3CH_2CH_2CH_2-$	Butyl	Bu–
$(CH_3)_2CH-$	Isopropyl	i-Pr–
$(CH_3)_2CHCH_2-$	Isobutyl	i-Bu–
$CH_3CH_2CHCH_3$ 		sec-Butyl	s-Bu–
$(CH_3)_3C-$	tert-Butyl	t-Bu–

IUPAC, International Union of Pure and Applied Chemistry.

In the naming of hydrocarbons, the substituents are alkyl groups which are derived from other alkanes, usually shown as R. Some common examples are shown in Table 2.2.

We will talk more about substitutive naming as needed to deal with other functional classes. Appendices 2 and 3 carry some additional details.

2.3 SIMPLE UNSATURATED HYDROCARBONS (ALKENES AND ALKYNES)

Compared with saturated hydrocarbons, alkenes and alkynes are chemically much more reactive because of the unsaturated (multiple bond) functional group. In multiple bonds, carbon bonds are in either sp^2 (C=C) or sp (C≡C) hybrid states, and take on trigonal or linear shapes. They may be in straight chain, branched, and cyclic forms.

We use the same naming rules as for alkanes, except that the ending of the root name is -ene (alkene) and -yne (alkyne). If there is any doubt, the atom number of the lower numbered carbon in the multiple bonds must be included in the name. Table 2.3 lists selected examples.

Some common unsaturated fragments are shown with their common and IUPAC names in Table 2.4.

Table 2.3 Selected Alkenes and Alkynes

IUPAC Name	Molecular Formula	Structural Formula	M.P. (°C)	B.P. (°C)
Ethene	C_2H_4	$H_2C{=}CH_2$	−169	−102
Propene	C_3H_6	$H_3CHC{=}CH_2$	−185	−48
1-Butene	C_4H_8	$H_3CH_2CHC{=}CH_2$	−185	−6.5
1-Pentene	C_5H_{10}	$H_3C(H_2C)_2HC{=}CH_2$	−138	30
1-Hexene	C_6H_{12}	$H_3C(H_2C)_3HC{=}CH_2$	−140	64
1-Octene	C_8H_{16}	$H_3C(H_2C)_5HC{=}CH_2$	−102	123
1-Decene	$C_{10}H_{20}$	$H_3C(H_2C)_7HC{=}CH_2$	−87	171
cis-2-Butene[a]	C_4H_8		−140	4
trans-2-Butene[a]	C_4H_8		−106	1
2-Methylpropene	C_4H_8	$(H_3C)_2C{=}CH_2$	−141	−7
Cyclopentene	C_5H_8		−93	46
Cyclohexene	C_6H_{10}		−104	83
Ethyne	C_2H_2	$HC{\equiv}CH$	−82	−75
Propyne	C_3H_4	$H_3CC{\equiv}CH$	−101	−23
1-Butyne	C_4H_6	$H_3CH_2CC{\equiv}CH$	−122	8
1-Pentyne	C_5H_8	$H_3C(H_2C)_2C{\equiv}CH$	−98	40
1-Hexyne	C_6H_{10}	$H_3C(H_2C)_3C{\equiv}CH$	−124	72
2-Hexyne	C_6H_{10}	$H_3C(H_2C)_2C{\equiv}CCH_3$	−92	84
3-Hexyne	C_6H_{10}	$H_3CH_2CC{\equiv}CCH_2CH_3$	−51	81

IUPAC, International Union of Pure and Applied Chemistry.
[a]The prefixes cis- and trans- are covered in Chapter 3.

Table 2.4 Selected Unsaturated Groupings

Group	Common Name	IUPAC Name
$H_2C{=}CH{-}$	Vinyl	Ethenyl
$H_2C{=}CHCH_2{-}$	Allyl	2-Propenyl
$HC{\equiv}C{-}$	–	Ethynyl
$HC{\equiv}CCH_2{-}$	Propargyl	2-Propynyl

IUPAC, International Union of Pure and Applied Chemistry.

2.4 COMPLEX UNSATURATED SYSTEMS (POLYENES AND AROMATICS)

There are many compounds in nature that have more than one multiple bond. These multiple bonds can have different relationships to each other. Because these arrangements can have an effect on structure and reactivity, it is important to classify the relationship between the multiple bonds in polyenes. Figure 2.5 shows the possible relationship that can exist between multiple bonds.

If there is more than one multiple bond, all the relevant location points must be shown in the name. The relevant ending is changed to -diene, -triene, -diyne, etc. as needed. If both C=C and C≡C are present, the ending becomes -enyne, and the chain numbering is chosen to give the set of lower numbers.

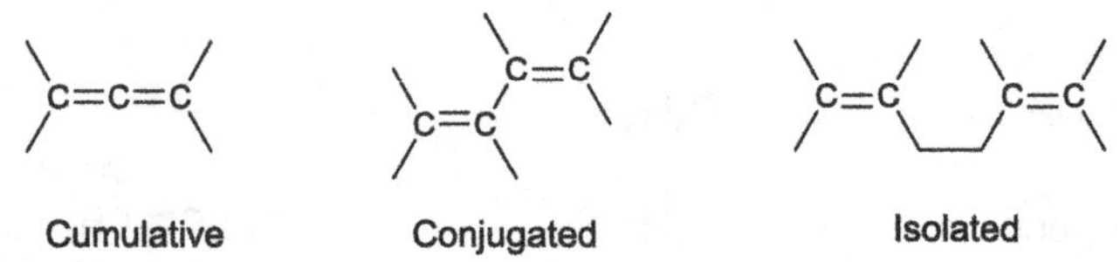

| Cumulative | Conjugated | Isolated |

FIGURE 2.5
Types of polyene systems.

Where two double bonds are directly connected, the system is called **cumulative**. To bond in this way, the central carbon must be *sp*-hybridized. This affects the shape of these molecules. If multiple bonds are separated by one single bond, the multiple bonds form a **conjugated** system. In these systems, the multiple bonds can have an electronic effect on each other. Finally, where multiple bonds are more than one single bond apart, they do not affect each other and act as **isolated** multiple bonds.

Aromatic compounds, or arenes, are a special class of conjugated polyenes. Their physical and chemical properties come from the special **delocalized** arrangement of their double bonds. This conjugated arrangement of alternate single and double bonds is further discussed in Chapter 4. As shown in Figure 2.6, the parent structure, benzene, may be drawn in a number of ways. In this book, we discuss only derivatives of benzene. This is enough to show the special nature of the compound class.

FIGURE 2.6
Common representations of benzene.

Table 2.5 shows the naming of examples of these systems. Most simple aromatics are named as derivatives of benzene. However, many historical common names are still used and can form the basis for certain IUPAC names.

As a substituent, benzene is usually written as C_6H_5- or phenyl (Ph–). We can shorten any arene to Ar–, the aromatic equivalent to the alkyl R– grouping.

Table 2.5   Selected Common Aromatics (Arenes)

Structural Formula	Common Name	IUPAC Name	M.P. (°C)	B.P. (°C)
	Benzene	Benzene	5	80
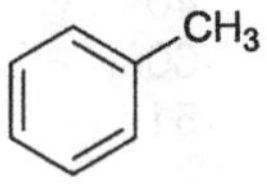	Toluene	Methylbenzene	−93	111
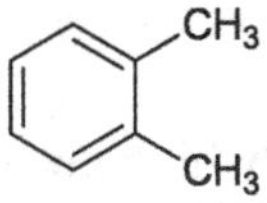	o-Xylene	1,2-dimethylbenzene	−25	143
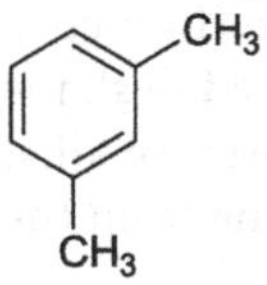	m-Xylene	1,3-dimethylbenzene	−48	138
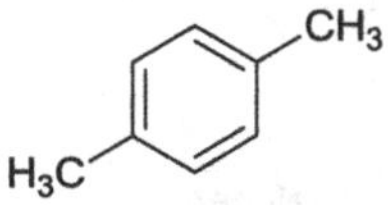	p-Xylene	1,4-dimethylbenzene	13	138
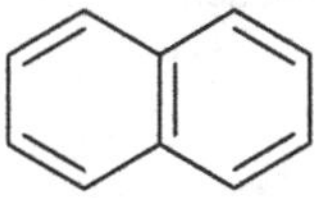	Naphthalene		80	218
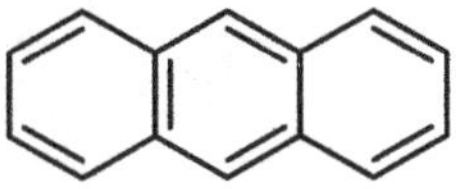	Anthracene		16	340
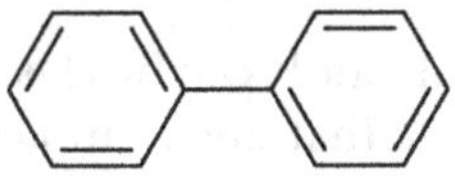	Biphenyl		69	255

IUPAC, International Union of Pure and Applied Chemistry.

The positions of substituents on the benzene ring can be shown by numbers. This is necessary when there are three or more substituents. In disubstituted benzenes, the relative substituent positions can also be given by the following prefixes. o- (ortho) to show a 1,2-, m- (meta) to show a 1,3-, and p- (para) to show a 1,4-substitution pattern.

Later sections in this chapter do not always give aromatic examples of the other functional classes. However, aromatic examples of all of these do exist and, in fact, are common.

2.5   ALKYL HALIDES

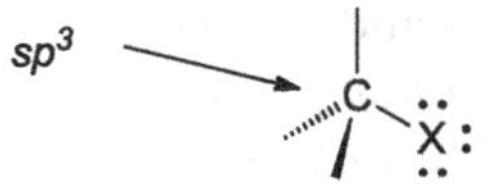

sp^3

X = F, Cl, Br, I

Formula	Common Name	IUPAC Name	B.P. (°C)
CH_3Cl	Methyl chloride	Chloromethane	−24
CH_3Br	Methyl bromide	Bromomethane	5
CH_3I	Methyl iodide	Iodomethane	42
CH_2Cl_2	Methylene chloride	Dichloromethane	40
$(CH_3)_2CHBr$	Isopropyl bromide	2-Bromopropane	60
$(CH_3)_3CCl$	tert-Butyl chloride	2-Chloro-2-methylpropane	51

Table 2.6 Selected Common Alkyl Halides

IUPAC, International Union of Pure and Applied Chemistry.

This is the first functional class that has a heteroatom. The relatively high electronegativity of the halogens gives a highly polar covalent bond (Inductive effect, Chapter 1). This does not change the sp^3 hybrid state or tetrahedral shape of the carbon, but it does give a reactive site that controls the chemistry of alkyl halides. Table 2.6 shows the IUPAC naming of alkyl halides that come from the hydrocarbon parents, with the halogen atom treated as a substituent.

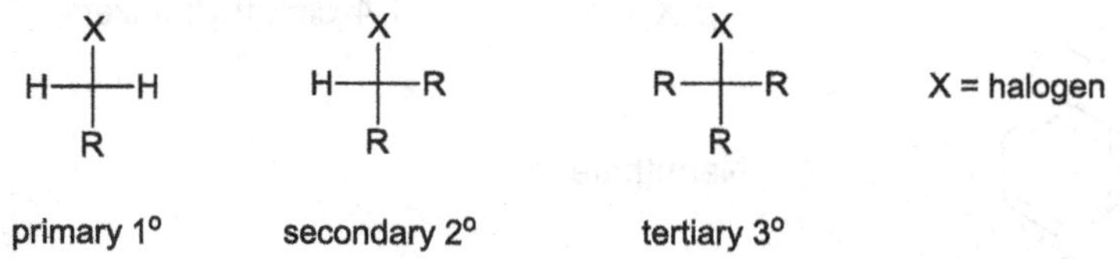

primary 1°   secondary 2°   tertiary 3°

FIGURE 2.7
Classification of 1°, 2°, and 3° alkyl halides.

Figure 2.7 shows that, similar to alkanes, alkyl halides and alcohols can be classified as 1° (primary), 2° (secondary), and 3° (tertiary). Note that the nominal oxidation number of the carbon bonded to the halogen changes from −1 in primary to +1 in tertiary. This change explains why there is a difference in reactivity across the range of alkyl halides.

2.6 ALCOHOLS, PHENOLS, ETHERS, AND THEIR SULFUR EQUIVALENTS (THIOLS AND THIOETHERS)

Figure 2.8 shows that all of these functional classes have a general structure in which carbon is connected by a single bond to one electronegative heteroatom. This gives a polar single bond between the heteroatom and the saturated sp^3-hybridized carbon. The nominal oxidation numbers are the same as for alkyl halides.

The heteroatom is also sp^3-hybridized and the tetrahedral shape of the functional group is well defined. It is common not to show the two lone pairs of electrons on the oxygen and sulfur. However, we must not forget the lone pairs,

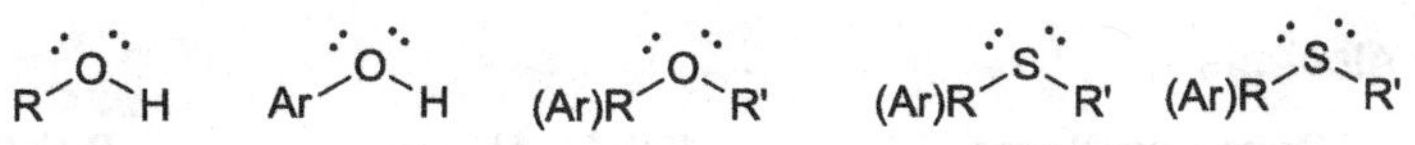

FIGURE 2.8
Structures of alcohols, phenols, ethers, thiols, and thioethers.

because they play an important role in the physical and chemical properties of these compounds.

Although alcohols and phenols have the same hydroxyl (–OH) functional group, their properties are very different. This is because of the different effect of the aromatic (–Ar) group in place of an alkyl (–R) group. In Chapter 6, we discuss the impact of this difference on hydroxyl acidity. In Chapter 8, we see the central role that these hydroxyls play in the structure and chemistry of biological molecules. We can think of alcohols and phenols as derivatives of a parent water molecule in which one of the H atoms is replaced by an alkyl or aryl group.

In ethers, the remaining H atom of alcohols and phenols is replaced by a second carbon substituent. The two carbon groups may be the same—symmetrical ethers, or different—unsymmetrical ethers. Cyclic examples that have the two ends of the same carbon chain linked by a common oxygen atom are also common.

Thiols and thioethers (sulfides) are the sulfur versions of the alcohols/phenols and ethers. The classifications of 1°, 2°, and 3° follow the pattern shown for alkyl halides in Figure 2.7.

The change in structure from alcohols to ethers causes large differences in their properties. These differences are related to the presence or absence of the highly polar hydroxyl group. The hydroxyl group can participate in hydrogen bonding, similar to that in water. As shown by Figure 2.9, hydrogen bonding has a large effect on physical properties such as boiling/melting points and solubilities. In Chapter 6, you will see that the chemical properties of acidity/basicity are also affected.

$CH_3CH_2CH_2CH_3$ $CH_3OCH_2CH_3$ $CH_3CH_2CH_2OH$

Butane   Methoxyethane   1-Propanol
Mol. wt. 58   Mol. wt. 60   Mol. wt. 60
B.p. -0.5°C   B.p. 8°C   B.p. 97

FIGURE 2.9
The effect of hydrogen bonding on boiling point.

2.6.1 Naming

The substitutive naming of these classes follows the general rules developed in Section 2.2.3. For alcohols, the ending -ol replaces the -e in the parent alkane, as listed in Table 2.7. If more than one hydroxyl group is present, the appropriate ending such as -diol or -triol is used in the name. The position of the hydroxyl

Table 2.7	Some Common Alcohols		
Formula	Common Name	IUPAC Name	B.P. (°C)
CH_3OH	Methyl alcohol	Methanol	65
CH_3CH_2OH	Ethyl alcohol	Ethanol	78
$CH_3CH_2CH_2OH$	n-Propyl alcohol	1-Propanol	97
$(CH_3)_2CHOH$	Isopropyl alcohol	2-Propanol	82
$(CH_3)_2CHCH_2OH$	Isobutyl alcohol	2-Methyl-1-propanol	108
$(CH_3)_3COH$	tert-Butyl alcohol	2-Methyl-2-propanol	82
$HOCH_2CH_2OH$	Ethylene glycol	1,2-Ethanediol	197
$HOCH_2CH_2(OH)CH_2OH$	Glycerol	1,2,3-Propanetriol	290
⬡—OH		Cyclohexanol	160
⬡—CH_2OH	Benzyl alcohol	Phenylmethanol	205

IUPAC, International Union of Pure and Applied Chemistry.

groups is given by chain numbers. If the alcohol is not the major functional group (Appendix 3), then the hydroxyl group is named as a hydroxy- substituent.

In phenols, the hydroxyl group is attached directly to an aromatic system (arene). They are usually named as substituted derivatives of the parent arene. However, as Table 2.8 shows, common names are often still used.

For ethers, there is no systematic ending for substitutive naming. In most simple cases their names are based on the longer chain parent backbone H–R', and then –OR is treated as a substituent. The name of the –OR group, which is an alcohol without its hydrogen, is a combination of the names of the alkyl –R group with -oxy to give the alkyloxy substituent. This is usually shortened to alkoxy when the carbon chain has five or less carbon atoms. For example, CH_3CH_2O- is ethoxy rather than ethyloxy. This is then added to the parent alkane –R' name as shown in the examples in Table 2.9.

Table 2.10 shows some thiols and thioethers, the sulfur equivalents of the alcohols and ethers. The thiols and thioethers are named using the ending -thiol and the class name sulfide. Appendix 3 shows another way to deal with these as substituents.

2.7 AMINES

In amines, the functional group is based on the amino group, $-NH_2$. The compound class can be seen as derivatives of ammonia, NH_3, with the hydrogen atoms replaced by carbon substituents. Therefore, as Figure 2.10 shows, they have the same sp^3 structure.

Figure 2.11 shows the classification into 1° (primary), 2° (secondary), 3° (tertiary), and 4° (quaternary) amines. These match the substitution of the

Table 2.8 Some Common Phenols

Formula	Common Name	IUPAC Name	M.P. (°C)
OH	Phenol	Phenol	41
OH, CH₃	o-Cresol	o-(or 2-) Methylphenol	31
OH, OH	Catechol	Benzene-1,2-diol	105
OH, HO	Resorcinol	Benzene-1,3-diol	110
HO, OH	Hydroquinone	Benzene-1,4-diol	172
NO₂, O₂N, OH, NO₂	Picric acid	2,4,6-Trinitrophenol	122
OH	α-Naphthol	Naphthalen-1-ol	96

IUPAC, International Union of Pure and Applied Chemistry.

hydrogens which are attached to the nitrogen, with the quaternary example equivalent to a protonated ammonium ion.

This quaternary amine shows the important role of the lone pair on nitrogen in the chemistry of the amines. The carbon substituent groups do not need be the same, but they are all bound to the central nitrogen by single σ-bonds. Note that alicyclic and aromatic amines are also relatively common.

Table 2.11 reveals that primary amines can be named by substitutive names, in which the systematic ending -amine replaces the -e of the parent chain. The carbon, to which the amino group is joined, is numbered. For secondary and tertiary amines, the name depends on whether the substituents are all same or

Table 2.9 Selected Common Ethers			
Formula	Common Name	IUPAC Name	B.P. (°C)
CH_3OCH_3	Dimethyl ether	Methoxymethane	−24
$CH_3OCH_2CH_3$	Ethyl methyl ether	Methoxyethane	7
$CH_3CH_2OCH_2CH_3$	Diethyl ether	Ethoxyethane	34
$(CH_3CH_2CH_2CH_2)_2O$	Butyl ether	Butoxybutane	142
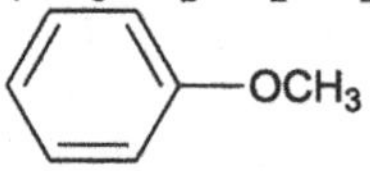 ⬡—OCH_3	Anisole	Methoxybenzene	154
 ◁O	Ethylene oxide	Oxirane	11
(tetrahydrofuran ring)	Tetrahydrofuran	Oxolane	67
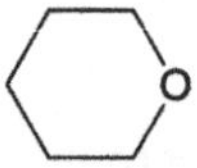 (tetrahydropyran ring)	Tetrahydropyran	Oxane	88

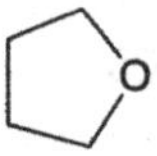

IUPAC, International Union of Pure and Applied Chemistry.

Table 2.10 Selected Thiols and Thioethers			
Formula	Common Name	IUPAC Name	B.P. (°C)
CH_3SH	Methyl mercaptan	Methanethiol	6
CH_3CH_2SH	Ethyl mercaptan	Ethanethiol	35
$CH_3CH_2CH_2SH$	n-Propyl mercaptan	1-Propanethiol	68
CH_3SCH_3	Dimethyl sulfide	Methylsulfanylmethane	38
$(ClCH_2CH_2)_2S$	(Mustard gas)	Bis(2-chloroethyl)sulfide	218

IUPAC, International Union of Pure and Applied Chemistry.

pyramidal sp^3 ⇌ trigonal planar sp^2 ⇌ pyramidal sp^3

FIGURE 2.10
Amine structure and inversion.

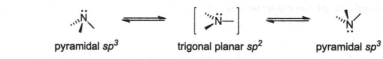

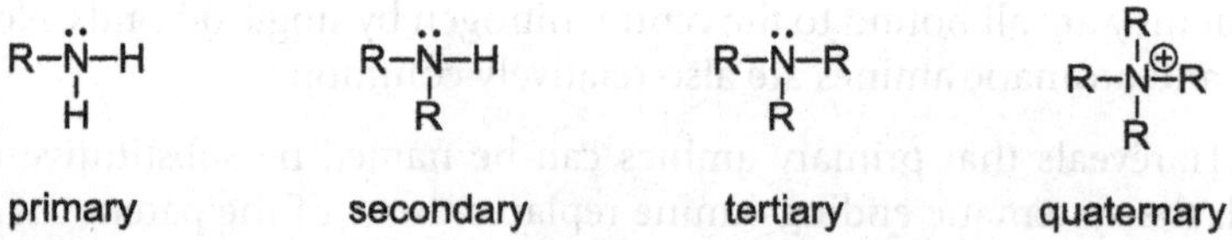

primary   secondary   tertiary   quaternary

FIGURE 2.11
Amines classification.

Table 2.11	Selected Amines	
Formula	**IUPAC Name**	**B.P. (°C)**
CH_3NH_2	Methanamine	–6
$(CH_3)_2NH$	N-Methylmethanamine	6
$(CH_3)_3N$	N,N-Dimethylmethanamine	3
$(CH_3)_2NCH_2CH_3$	N,N-Dimethylethylamine	36
⬡–NH_2	Phenylamine	184
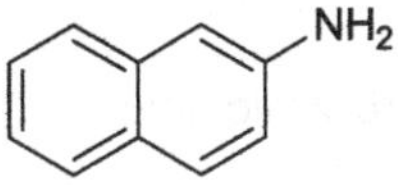 ⬡⬡–NH_2	2-Aminonaphthalene	306
⬡N	Pyridine	115
⬠NH	Pyrrolidine	87
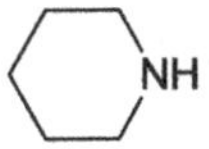 ⬡NH	Piperidine	106

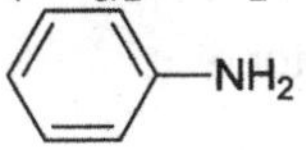

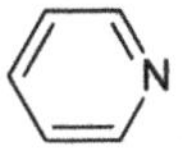

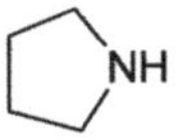

IUPAC, International Union of Pure and Applied Chemistry.

different. Appendix 4 has additional examples. Finally, if a functional group of higher priority is present, then the $-NH_2$ is treated as an -amino substituent.

2.8   COMPOUNDS WITH CARBONYL GROUPS

Table 2.12 lists several classes of organic compounds that have the important structural feature called a **carbonyl group**. As Figure 2.12 shows, a carbonyl group has a carbon with a double bond to oxygen. This functional group has planar geometry because of the sp^2-hybridized carbon and oxygen. The C=O double bond is also shown as highly polar. This can be seen as the dipolar combination **III** of the extreme forms **I** and **II**. In Chapter 4, this concept is discussed in more detail. The combination of shape and polarity has a major effect on the structure, properties, and reactivity of these compounds.

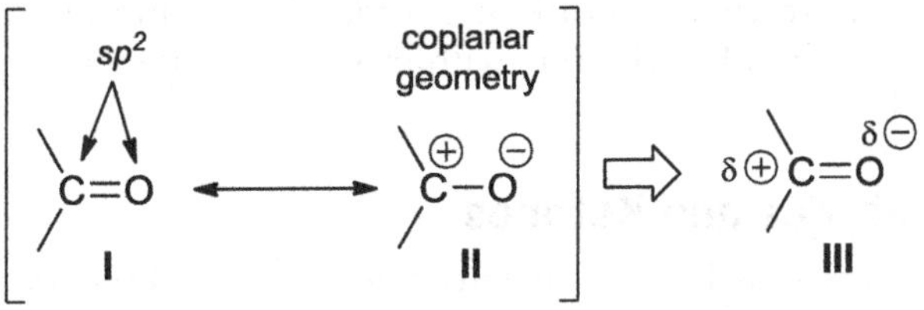

FIGURE 2.12
Structural features of the carbonyl group.

Table 2.12 Some Classes of Carbonyl Compounds		
General Structure	**Carbonyl Description**	**Compound Class**
R—C=O, H (R, H on carbonyl)	Carbonyl	Aldehyde
R—C=O, R (R, R on carbonyl)	Carbonyl	Ketone
R—C=O, HO	Carboxyl	Carboxylic acid
R—C=O, OR'	Acyl	Ester
R—C=O, H₂N	Acyl	Amide
R—C=O, F/Cl/Br/I	Acyl	Acid (acyl) halide
R—C(=O)—O—C(=O)—R	Acyl	Acid anhydride

In the study of organic functional groups it is useful to think of the compound classes as two sets of parallel structures. These are related by the absence or presence of a carbonyl group. Using this approach, the set of functional classes in following text are simply a repeat of the set already shown earlier. The difference is simply the presence of a carbonyl group.

Therefore, aldehydes and ketones are carbonyl-modified hydrocarbons. Carboxylic acids are parallel to alcohols, and acyl halides, esters, and amides are the carbonyl equivalents to halides, ethers, and amines, respectively. One further class, anhydrides, comes from the ether equivalent if both carbon atoms of the C–O–C bond are modified to carbonyl groups.

The properties and reactivity of carbonyl compounds is mostly a combination of the features of the carbonyl group with those of the functional group that is being modified.

2.8.1   Aldehydes and Ketones

Aldehydes or ketones can be separated from other carbonyl classes of compound on the basis of the number of bonds to heteroatoms. This affects the nominal oxidation number of the functional group carbon.

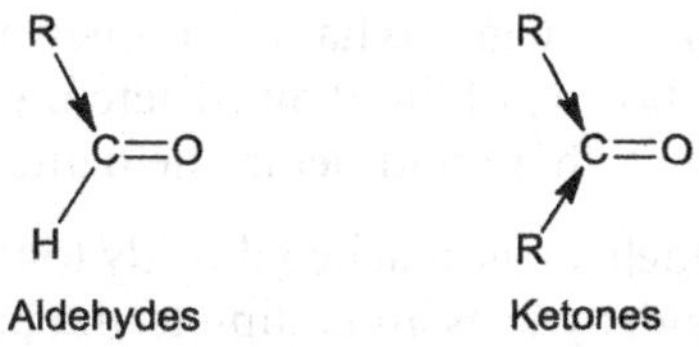

FIGURE 2.13
Structural differences between aldehydes and ketones.

Aldehydes and ketones have carbonyl carbon atoms with nominal oxidation numbers of +1 and +2. Because of this, the properties of these classes depend mainly on the carbonyl group. Any further difference between aldehydes and ketones is because of the different number of carbon attachments on the carbonyl carbon. As shown in Figure 2.13, the overall inductive effect on the carbonyl group in the two compound classes is different. Because this determines how polar the carbonyl bond is, it affects the chemical reactivity of the group.

Table 2.13	Selected Aldehydes and Ketones		
Formula	**Common Name**	**IUPAC Name**	**B.P. (°C)**
HCHO	Formaldehyde	Methanal	−21
CH_3CHO	Acetaldehyde	Ethanal	21
CH_3CH_2CHO	Propionaldehyde	Propanal	49
$CH_3(CH_2)_2CHO$	Butyraldehyde	Butanal	76
$CH_2{=}CHCHO$	Acrolein	Propenal	53
$CH_3CH{=}CHCHO$	Crotonaldehyde	2-Butenal	104
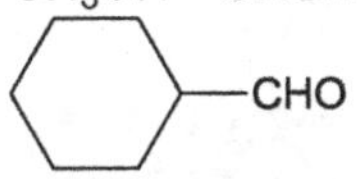 CHO	Formylcyclohexane	Cyclohexanecarbaldehyde	161
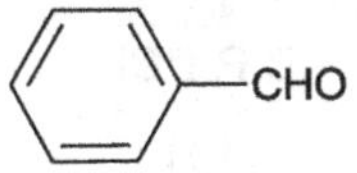 CHO	Benzaldehyde	Benzaldehyde	178
$CH_3CO{\cdot}CH_3$	Acetone	Propanone	56
$CH_3CO{\cdot}CH_2CH_3$	Ethyl methyl ketone	Butanone	80
$CH_3CH_2CO{\cdot}CH_2CH_3$	Diethyl ketone	3-Pentanone	102
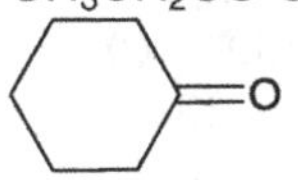	–	Cyclohexanone	155
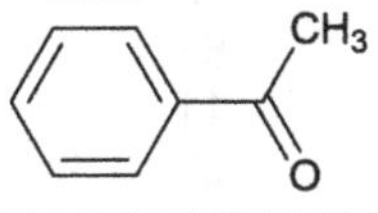	Methyl phenyl ketone	Acetophenone	202

IUPAC, International Union of Pure and Applied Chemistry.

2.8.1.1 NAMING

For convenience, aldehydes are often written in the short form as R–CHO and ketones as R–CO–R′. General substitutive naming is done by replacing the -e of the parent chain with -al (aldehydes) and -one (ketones).

Because the aldehyde functional group has a hydrogen atom as one substituent, it must be at the end (carbon 1) of the chain. Therefore, as Table 2.13 shows, it is not necessary to include a chain number in the name.

In cyclic examples, the aldehyde is attached directly to the ring. Here the substitutive naming uses carbaldehyde as an ending to the parent ring name. Sometimes it is necessary to name the aldehyde or ketone group as a substituent. In these cases, the prefixes formyl- and oxo- are used along with the chain number.

2.8.2 Carboxylic Acids

The carboxyl functional group can be seen as a combination of the carbonyl and hydroxyl functionalities. It is often written as the short forms R–CO_2H or R–COOH. Chapter 6 shows how the properties of the carbonyl and hydroxyl groups combine to give compounds with special acidic properties. The sp^2 carboxyl carbon has three bonds to oxygen and a nominal oxidation number of +3. However, the single carbon example of methanoic acid is an exception. It still has a hydrogen substituent and therefore a carbon nominal oxidation number of +2.

Several IUPAC-approved names are used for carboxylic acids. Table 2.14 gives examples of IUPAC and substitutive naming. The last -e in the parent chain changes to -oic, and this is written before the word acid. In substituted examples, numbering starts from the carboxyl functional group.

If the carboxylic group is directly attached to a ring, the naming is done by adding the ending -carboxylic acid to the parent ring name. There are many common names that are still widely used, for example formic and acetic acids.

Table 2.14   Selected Common Carboxylic Acids

Structural Formula	Common Name	IUPAC Name	B.P. (°C)
HCO_2H	Formic acid	Methanoic acid	101
CH_3CO_2H	Acetic acid	Ethanoic acid	118
$CH_3CH_2CO_2H$	Propionic acid	Propanoic acid	141
$CH_3(CH_2)_2CO_2H$	Butyric acid	Butanoic acid	164
			M.P. (°C)
$CH_2{=}CHCO_2H$	Acrylic acid	Propenoic acid	13
$CH_3CH{=}CHCO_2H$	Crotonic acid	2-Butenoic acid	72
$ClCH_2CO_2H$	Chloroacetic acid	Chloroethanoic acid	63
Cl_2CHCO_2H	Dichloroacetic acid	Dichloroethanoic acid	11
Cl_3CCO_2H	Trichloroacetic acid	Trichloroethanoic acid	58
⬡–CO_2H	–	Cyclohexanecarboxylic acid	30
⬡–CO_2H	Benzoic acid	Benzoic acid	122

IUPAC, International Union of Pure and Applied Chemistry.

2.8.3 Carboxylic Acid (Acyl) Derivatives

The classes of carboxylic acid derivatives are also modifications of the carbonyl functional group. All of them have the common R–CO– **acyl** fragment. These fragments are made up of any carbon group attached to a carbonyl group. Because of their chemical relationships as seen in Table 2.15, they are seen as derivatives of carboxylic acids.

Table 2.15 Some Common Acyl Root Names		
Parent Acid	**Acyl Group**	**Name**
HCO_2H	$HCO-$	Formyl (methanoyl)
CH_3CO_2H	CH_3CO-	Acetyl (ethanoyl)
$CH_3CH_2CO_2H$	CH_3CH_2CO-	Propanoyl
$CH_2{=}CHCO_2H$	$CH_2{=}CHCO-$	Acryloyl (propenoyl)
		Benzoyl

Like acids, the carbon of the functional group has three bonds to heteroatoms and a nominal oxidation number of +3. Carboxylic acids can be seen as an alcohol hydroxyl group modified by the carbonyl function. Table 2.16 shows that acyl halides, esters, and amides are carbonyl modified versions of organic halides, ethers, and amines.

Table 2.16 Selected Acyl-Based Compounds	
Acyl Compound	**Name**
CH_3COCl	Acetyl (ethanoyl) chloride
	Benzoyl chloride
$CH_3CO_2CH_2CH_3$	Ethyl acetate (ethanoate)
$CH_2{=}CHCO_2CH_3$	Methyl acrylate (propenoate)
$CH_3(CH_2)_4CONH_2$	Hexanamide
	Cyclopentanecarboxamide
$CH_3CH_2CONHCH_3$	N-Methylpropanamide
$CH_3CO-O-COCH_3$	Acetic (ethanoic) anhydride
	Benzoic anhydride

2.8.3.1   NAMING

Acyl halides (RCO–Halogen) are given two-word functional class names. The corresponding acyl group comes from the parent acid by replacing the terminal -ic with -yl. Then this is written before the appropriate halide. For example, $CH_3CH_2CH_2COBr$ is butanoyl bromide.

Esters (RCO–OR′) are given two-word names in a similar way to the naming of salts. The R′ group becomes the first word, and the second word is formed by changing the parent acid -ic to -ate. For example, $CH_3CH_2CH_2CO_2CH_3$ is methyl butanoate.

Amides (RCO–NH$_2$) are named by replacing the name of the corresponding acid by the systematic ending -amide. As with amines, the categories 1°, 2°, and 3° may exist for amides, and the naming is done in the same way. For example, $CH_3CH_2CH_2CONH_2$ is butanamide.

Acid anhydrides (RCO–O–COR′) are equal to two molecules of carboxylic acid which have combined with the loss of a water molecule. Symmetrical examples are named by replacing acid with anhydride in the parent carboxylic acid. For example, $CH_3CH_2CO–O–COCH_2CH_3$ is propionic anhydride.

2.8.3.2   NITRILES (CYANIDES)

Although nitriles do not have a carbonyl group, they are related chemically to carboxylic acids. Chapter 7 discusses this chemistry in more detail. The nitrile group (–C≡N) has carbon as an *sp*-hybrid because of the triple bond to the heteroatom. Therefore, as is clear from Figure 2.14, the carbon has the same formal oxidation number of +3 as the other acyl derivatives.

FIGURE 2.14
Structure and naming of nitriles (cyano derivatives).

The molecular formula of the nitrile group clearly shows that it equals an amide that has lost a water molecule. Simple members of the class are named by adding the ending -nitrile to the parent chain name, and the nitrile carbon is numbered as 1. More complex examples are named as derivatives of the corresponding carboxylic acids by changing the -ic to -onitrile, or by replacing the -carboxylic acid ending with -carbonitrile.

QUESTIONS AND PROGRAMS

Q 2.1. Write out at least three members of each of the homologous series that fit the following descriptions.
  (a) Unbranched acyclic carboxylic acids
  (b) Aliphatic terminal alkynes
  (c) Methyl ketones

PROGRAM 4 Structural Diagrams

A   As seen in Section 2.2.1, organic chemistry has different ways to show molecular structures. Which one do you use? This depends on the information that is needed or what is most convenient. Therefore, the user must often make a choice.

Study the fully extended structure of the molecule below and make a list of all the important features that you can find.

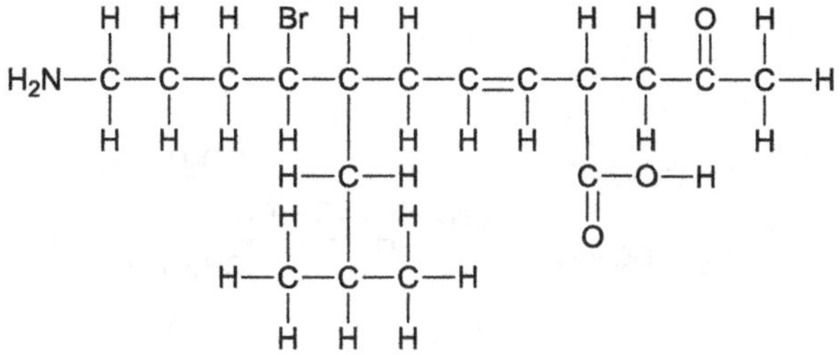

B   You should have quickly identified the functional groups: ketone, alkene, amine, carboxylic acid, and alkyl halide.

The extended structural diagram is not convenient and does not give information about the 3-D shape. However, it does show every bond in the molecule and makes sure that all valencies are correct.

Because molecules can be drawn in different ways, the fully extended form is not convenient for making comparisons between diagrams. Two of many different forms that can be drawn are:

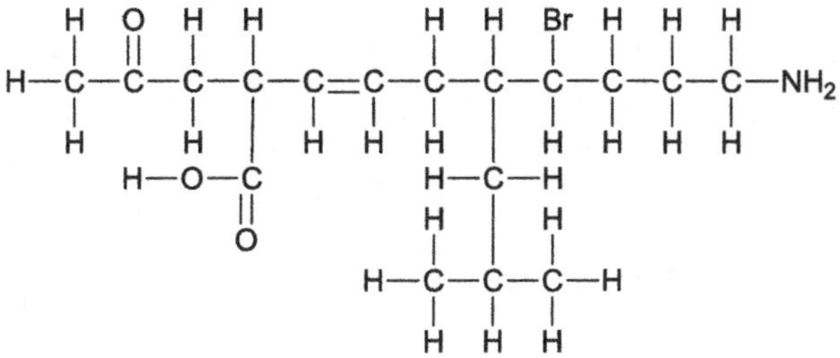

Continued...

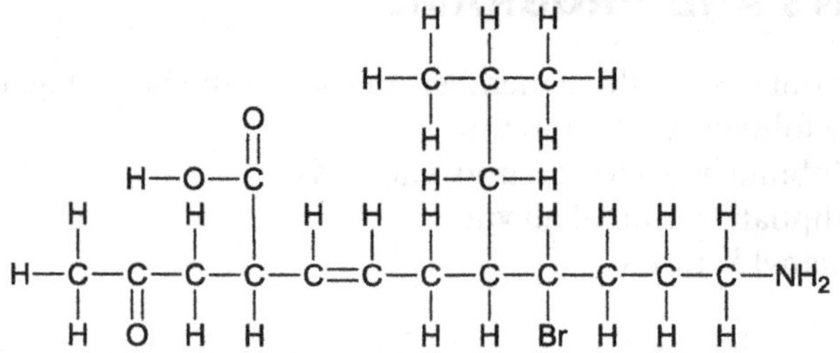

Draw a condensed representation that shows the important structural information.

C When you draw a structure in condensed form, you can use shorthand for parts of it and not others. This depends on exactly what you need the structure to show. For example, you might have drawn:

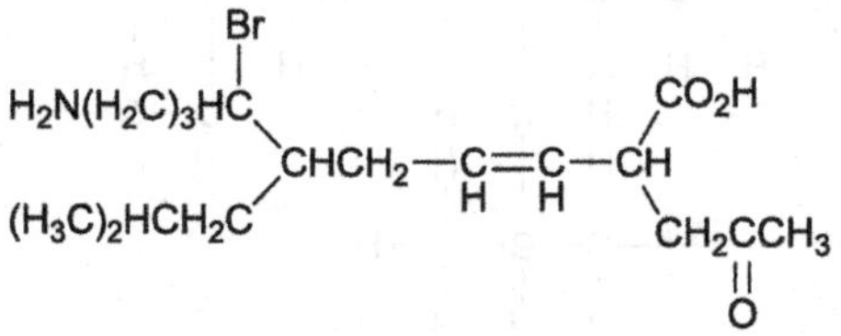

$H_2N(H_2C)_3HC$ — Br — CHCH$_2$—C=C—CH ... CO_2H
(H$_3$C)$_2$HCH$_2$C ... CH$_2$CCH$_3$

This condensed structure has a combination of bond lines and contracted group forms. All the functional groups are clearly shown, but there is no detail of molecular shape. In particular, this drawing does not show the geometry of the alkene C=C. This geometry can be very important for the identity of the molecule. There are many ways of drawing this mixed condensed form.

The most contracted form is the bond line structural diagram. In this form, both carbon and hydrogen atom labels are not shown. Instead they are shown as bends in the bond chain. Each bend is a carbon with the correct number of hydrogens to fill its valency. All other heteroatoms and attached hydrogen atoms are labeled.

Try to draw a bond line representation for the molecule.

D

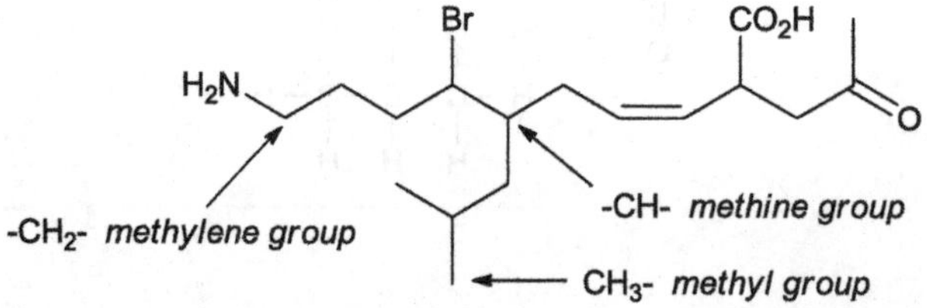

H_2N · · · Br · · · CO_2H

-CH$_2$- *methylene group*

-CH- *methine group*

CH$_3$- *methyl group*

Q 2.2. Study the molecular representations below and then convert them to the appropriate bond line structure equivalents.

$$CH_3CH_2CH_2CH(CH_3)CH_2CH_2CH_2C(OH)(Br)CH_2CH_3$$

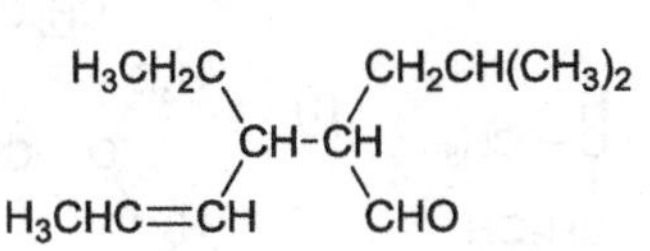

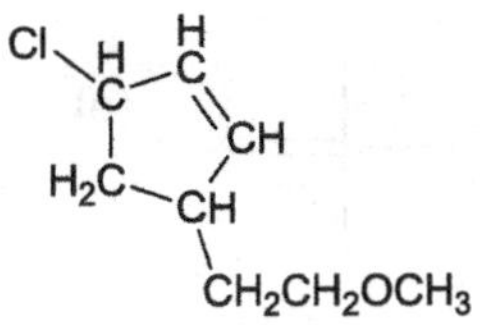

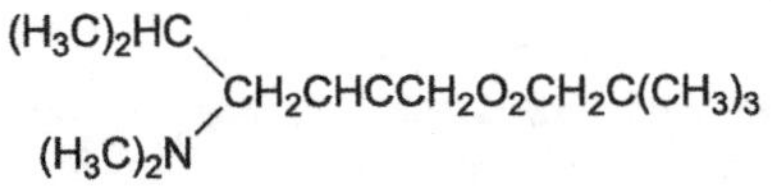

Q 2.3. Study the following molecular representations and then: (1) write the molecular formula for each; (2) sort the formulae into groups with identical molecular formula; (3) within each group from (2), identify any sets of identical molecules.

(a) $CH_3CH_2CH_2CH_2OH$

(b) $HOCH_2CH_2CH_2OH$

(c) $HOCH_2CH_2CH_2CH_3$

(d) $CH_3CH_2CHCH_3$

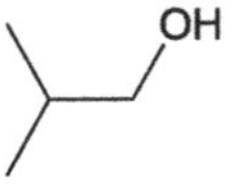

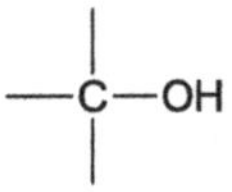

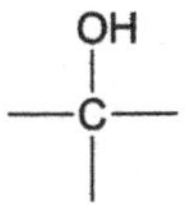

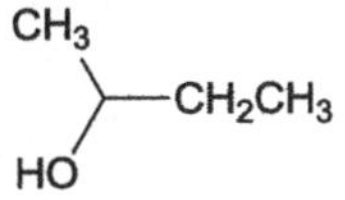

(j) $CH_3CH(OH)CH_2CH_3$

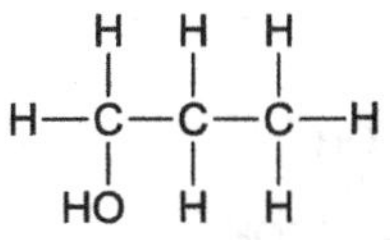

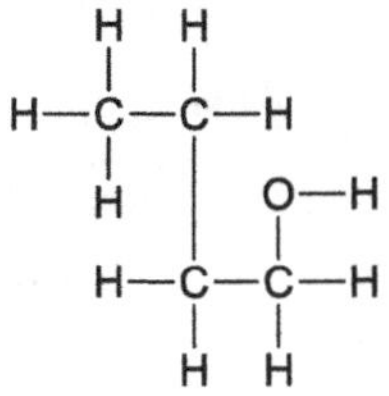

(m) $(CH_3)_3COH$

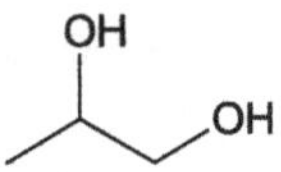

Q 2.4. For the following molecules, classify all the carbon centers as either primary (1°), secondary (2°), tertiary (3°), or quaternary (4°).

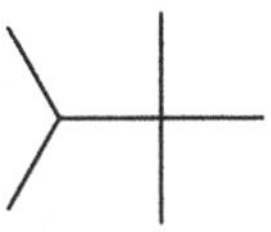

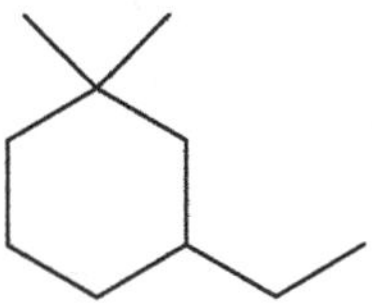

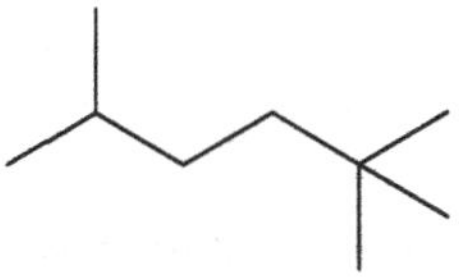

Q 2.5. Study the following structures. Mark any that are incorrect and show why this is so. Draw a correct structure.

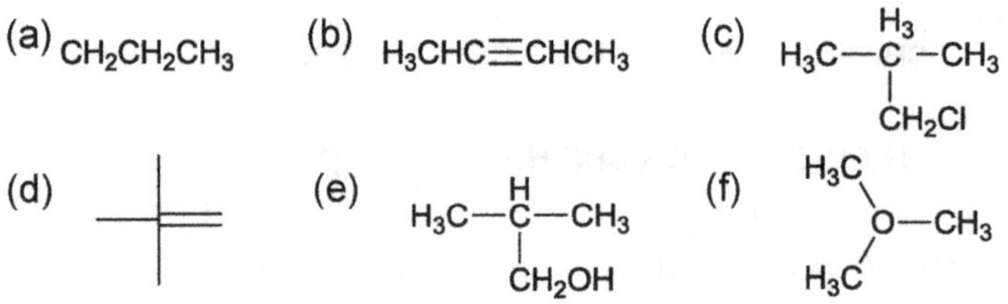

(a) $CH_2CH_2CH_3$

(b) $H_3CHC{\equiv}CHCH_3$

(c) $H_3C-\overset{\overset{\displaystyle H_3}{|}}{\underset{\underset{\displaystyle CH_2Cl}{|}}{C}}-CH_3$

(d) ✛ (structure drawing)

(e) $H_3C-\overset{\overset{\displaystyle H}{|}}{\underset{\underset{\displaystyle CH_2OH}{|}}{C}}-CH_3$

(f) $\begin{matrix} H_3C \\ \\ H_3C \end{matrix}\!\!\diagdown\!\!\underset{\diagup}{O}-CH_3$

PROGRAM 5 Carbon Oxidation Numbers

A Nominal oxidation numbers of carbon centers show relative electronegativity. It also shows the similarity or difference of oxidation states between the various carbon centers.

As shown in Section 2.2.2, all C–C bonds are seen as non-polar. Bonds between carbon and hydrogen are polarized toward carbon. Each of these then adds +1 to the oxidation state of the carbon. Bonds between carbon and common heteroatoms (O, N, S, Halogens) are all polarized toward the more electronegative heteroatoms. Therefore these bonds add −1 to the oxidation state of the carbon.

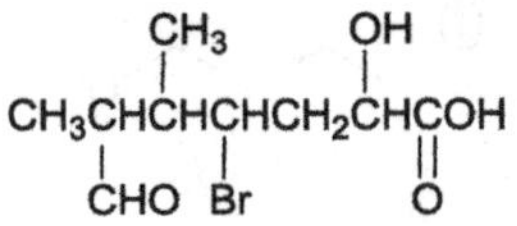

$$CH_3CH\underset{\underset{\displaystyle CHO\ Br}{|\qquad|}}{\overset{\overset{\displaystyle CH_3\qquad OH}{|\qquad\quad|}}{CHCHCH_2CHCOH}}$$

🖎 Give nominal oxidation numbers to the carbon centers in the above molecule.

B You should show the following:

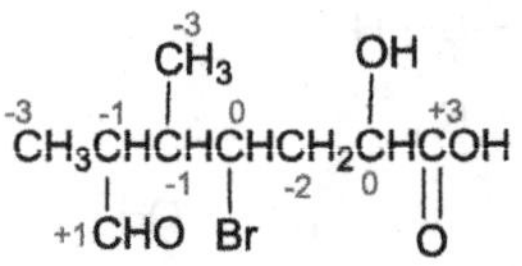

$$\underset{\underset{\displaystyle +1\,CHO\ \ Br}{|\ \ -1\ \ |\ \ -2\ \ 0\ ||}}{\overset{\overset{\displaystyle ^{-3}CH_3\qquad OH}{|\qquad\quad|}}{^{-3}CH_3\,^{-1}CHCHCH_2CHCOH}}$$

Notice the equivalence of certain centers. This means that they have the same oxidation state. Chapter 7 shows the full meaning of this in terms of functional group reactions.

Note also the equal oxidation states of alkyl halide and alcohol functional groups. The diagram also clearly shows the increase in oxidation number going from alcohol, to aldehyde, to carboxylic acid.

Q 2.6. Use Program 5 as a guide to give nominal oxidation numbers to carbon centers in the molecules in **Q 2.2**.

Q 2.7. Draw 3-D stereochemical diagrams of the *bold* carbon centers in the following molecules.

CH_2Cl_2 CH_3CH_2OH $CH_2{=}CHCH_2Br$

PROGRAM 6 IUPAC Naming

A IUPAC naming follows a series of steps in order to make a unique compound name. These names have the general form:

Prefix	**Parent**	**Suffix**
⇑	⇑	⇑
What and where are the substituents	How many carbons in the longest chain	What family of compound

🖎 Study the following molecule and develop the IUPAC name.

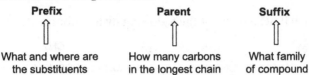

B Because the molecule consists of only saturated hydrocarbon portions, the family name is alkane. The longest continuous carbon chain is seven carbons long. This gives the molecular root as hept- and the molecule as a heptane.

$$H_3C$$
$$H_2C{-}CH_3$$
$$CH_3{-}CH{-}CH_2{-}CH{-}CH{-}CH_2{-}CH_3$$
$$CH_3{-}CH_2$$

🖎 Now give numbers to the backbone.

C You should have drawn.

$$H_3C \qquad H_2C{-}CH_3 \qquad \text{Branch points 2,4,5}$$
$$CH_3{-}CH{-}CH_2{-}CH{-}CH{-}CH_2{-}CH_3$$
$$\;\;1 \quad\;\; 2 \quad\;\; 3 \quad\; |4 \;\; 5 \quad\;\; 6 \quad\;\; 7$$
$$CH_3{-}CH_2$$

and not

$$H_3C \qquad H_2C{-}CH_3 \qquad \text{Branch points 3,4,6}$$
$$CH_3{-}CH{-}CH_2{-}CH{-}CH{-}CH_2{-}CH_3$$
$$\;\;7 \quad\;\; 6 \quad\;\; 5 \quad\; |4 \;\; 3 \quad\;\; 2 \quad\;\; 1$$
$$CH_3{-}CH_2$$

🖎 Identify the substituents and combine the information to give the name.

> **D** The alkyl substituents are a methyl group at C_2 and two ethyl groups at C_4 and C_5. List the substituents in alphabetical order and the IUPAC name can be generated.
>
> 4,5-Diethyl-2-methylheptane
>
> Note that numbers are separated by commas, and number and letters are separated by hyphens. Finally, the final substituent and the parent name are written as one word.

Q 2.8. Write IUPAC names for the following hydrocarbons.

(a)
$$CH_3$$
$$CH_3CH_2CH_2CHCHCH_3$$
$$CH_3$$

(b)
$$CH_3$$
$$CH_3CH_2CH_2CHCHCH_3$$
$$CH_2CH_2CH_2CH_3$$

(c)
$$CH_3 \quad CH_2CH_3$$
$$CH_3CHCH_2CCH_3$$
$$CH_2CH_3$$

(d)
$$CH_2CH_3$$
$$CH_2{=}CHCCH_2CH_3$$
$$CH_2CH_3$$

(e)
$$CH_3$$
$$H_3CHC{=}CHCHCH_2CH_3$$

(f)
$$CH_3$$
$$HC{\equiv}CCH_2CHCHCH_3$$
$$CH_3$$

Q 2.9. Draw structural formulae for the following.

(a) 2,4,4-Trimethylheptane   (b) 4-Ethyl-2-methylhexane
(c) 3-Ethyl-1-heptyne   (d) 3,5-Dimethyl-4-hexen-1-yne
(e) 1,1-Dimethylcyclopentane   (f) 1-Methyl-1,3-cyclopentadiene
(g) Triphenylmethane   (h) o-Hexylisobutylbenzene
(i) 1-Ethyl-4-nitrobenzene   (j) 2-Chloro-1,3,5-trinitrobenzene

Q 2.10. Draw structures for compounds that have the following features.
(a) A quaternary carbon and a secondary amine
(b) One secondary and two tertiary carbons
(c) An aldehyde bonded to a secondary carbon
(d) Two methylene and three methyl groups
(e) An isopropyl and a tertiary butyl group
(f) An alkene and a nitrile
(g) A secondary and a tertiary alkyl halide
(h) A dimethylphenyl group and a carboxylic acid
(i) An unsymmetrical ether
(j) A cyclic ester
(k) An alkyl-aryl ketone
(l) Non-aromatic with four coplanar carbon atoms.

Q 2.11. Classify each of the following functionalized compounds as either primary (1°), secondary (2°), tertiary (3°), or quaternary (4°).

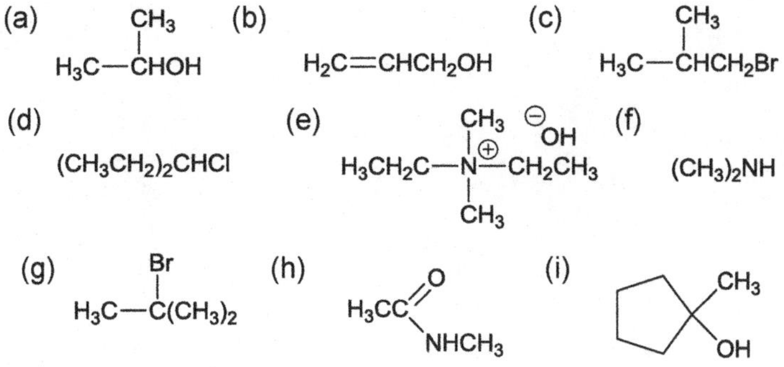

(a)
$$CH_3$$
$$H_3C—CHOH$$

(b)
$$H_2C\!=\!CHCH_2OH$$

(c)
$$CH_3$$
$$H_3C—CHCH_2Br$$

(d)
$$(CH_3CH_2)_2CHCl$$

(e)
$$CH_3 \quad ^\ominus OH$$
$$H_3CH_2C—\overset{\oplus}{N}—CH_2CH_3$$
$$CH_3$$

(f)
$$(CH_3)_2NH$$

(g)
$$Br$$
$$H_3C—C(CH_3)_2$$

(h)
$$H_3CC\overset{O}{\diagdown}NHCH_3$$

(i)

Q 2.12. Provide IUPAC names for the compounds in **Q 2.10**.

Q 2.13. Explain why each of the following names is wrong and give a correct IUPAC version.

(a) 1,3-dimethylbutane

(b) 4-methylpentane

(c) 2-ethyl-2-methylpropane

(d) 2-ethyl-3-methylpentane

(e) 4,4-dimethylhexane

(f) 2-propylpentane

(g) 5-butyloctane

(h) 4-isopropylbutane

Q 2.14. Give IUPAC names for the following compounds

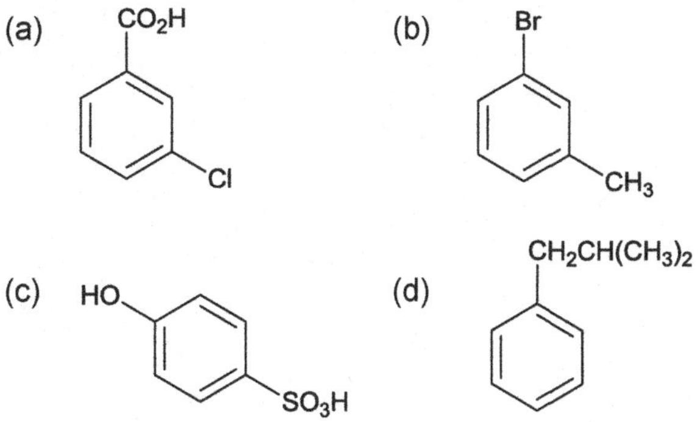

(a)
CO$_2$H
Cl

(b)
Br
CH$_3$

(c)
HO
SO$_3$H

(d)
CH$_2$CH(CH$_3$)$_2$

CHAPTER 3

Isomers and Stereochemistry

3.1 WHAT ARE ISOMERS?

Chapters 1 and 2 show that there are many ways to bond together the atoms of a molecular formula. **Isomers** are molecules with the same molecular formula, but with the atoms arranged differently. It is important to understand the relationships between structures that are isomers. There are different types of relationships, depending on the way the structures are put together.

If we know the causes of various types of isomers, we can understand the differences in physical and chemical properties between these organic compounds. The same simple ideas extend to the more complex systems found in nature.

3.2 STRUCTURAL ISOMERS

Structural isomers occur when the atoms of the same molecular formula are joined in different orders. This gives structurally different molecules that may have similar physical and chemical properties, Figure 3.1(a), or not, Figure 3.1(b).

To interconvert these isomers, relatively high energy is needed to break and make a number of strong single σ bonds, C–C ~348 and C–H ~415 kJ/mol. This is not usually practical.

(a)

	Butane	2-Methylpropane
	$CH_3CH_2CH_2CH_3$	$CH_3CH(CH_3)_2$
B.p.(°C)	-0.5	-12
	non-polar	non-polar

(b)

	Ethanol	Methyl ether
	CH_3CH_2OH	CH_3OCH_3
B.p.(°C)	78	-24
	highly polar	slightly polar

FIGURE 3.1
Structural isomers.

Organic Chemistry Concepts: An EFL Approach. http://dx.doi.org/10.1016/B978-0-12-801699-2.00003-1

3.3 CONFORMATIONAL ISOMERS

Conformational isomers, or conformers for short, are caused by the rotation around covalent single σ bonds and the three-dimensional (3-D) tetrahedral shape of the sp^3-hybridized centers. As you will see in Chapter 8, in larger biomolecules such as proteins and enzymes, the overall conformational shape of the molecule can be necessary for its biological activity.

3.3.1 Conformations in Alkanes

Two common types of diagrams are used to show conformers. See Figure 3.2. The first is a Sawhorse representation, which is an angled view along the rotating bond. The second is a Newman projection, which is an end-on view along the rotation bond with a circle to represent the front carbon center. Bonding to the front carbon is drawn to the center of the circle. Bonding to the rear carbon are drawn only to the edge of the circle.

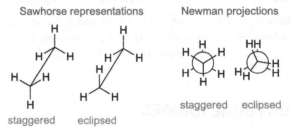

FIGURE 3.2
Extreme conformers of ethane.

Unlike structural isomers, conformers can interconvert easily because the energy for rotation is small compared to the energy in the system under normal conditions. Still, there are small energy differences between conformers, and some are more stable (of lower energy) than others.

For example, in ethane, the energy difference between the two extreme conformers is about 12 kJ/mol. The lower energy **staggered** conformer has the C–H bonds rotated as far apart as possible. The higher energy **eclipsed** conformer has the C–H bonds lined up as shown in Figure 3.2. Because 60–80 kJ is available at room temperature, interconversion over all the possible conformers between these two extremes occurs easily.

As seen in Figure 3.3, rotation around the central C–C bond for butane gives two different staggered forms and two different eclipsed forms. This is because two of the hydrogen substituents from the ethane model in Figure 3.2 are now methyl groups.

The simple situations in ethane and butane are multiplied as the molecules become more complex. In these, the specific rotational energies around each C–C bond can be different. As Figure 3.4 indicates, the lowest energy conformation in alkane systems usually tries to have the greatest number of bonds with staggered arrangement.

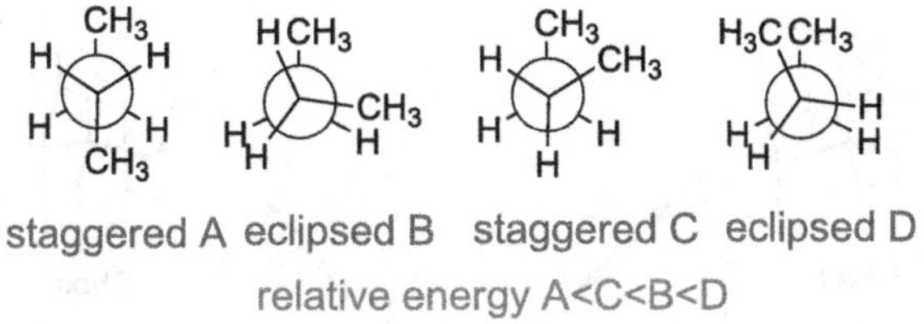

staggered A eclipsed B   staggered C eclipsed D

relative energy A<C<B<D

FIGURE 3.3
Staggered and eclipsed conformers for butane.

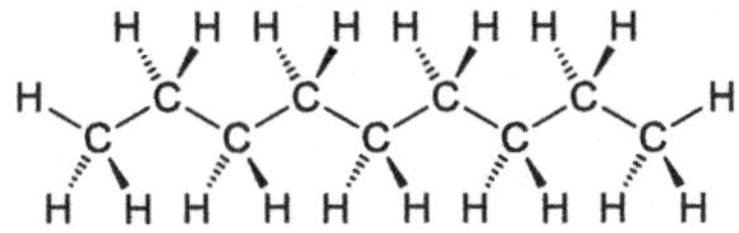

FIGURE 3.4
Fully extended staggered alkane array.

3.3.2 Conformations in Cycloalkanes

While $C_3–C_5$ cycloalkanes are nearly planar, rings of six or more carbons have enough flexibility for some rotation around the bonds. Because the bonded centers are sp^3-hybridized, this leads to conformational changes.

The very common cyclohexane ring shows this feature clearly. There are two extreme conformations that differ by 23 kJ/mol in energy. The lower energy conformer is called a chair, and the higher energy conformer is called a boat. See Figure 3.5.

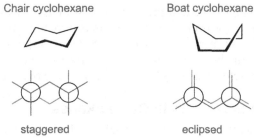

Chair cyclohexane   Boat cyclohexane

staggered   eclipsed

FIGURE 3.5
Cyclohexane conformers.

Cyclohexane can easily convert between the two possible lowest energy chair forms. This conversion goes through a boat, and needs about 45 kJ/mol to occur. See Figure. 3.6.

Note that the hydrogen substituents of the chair are divided into two sets of six each. The set in the overall plane of the ring is called equatorial. The other set, at right angles to the overall plane of the ring, is called axial. In unsubstituted

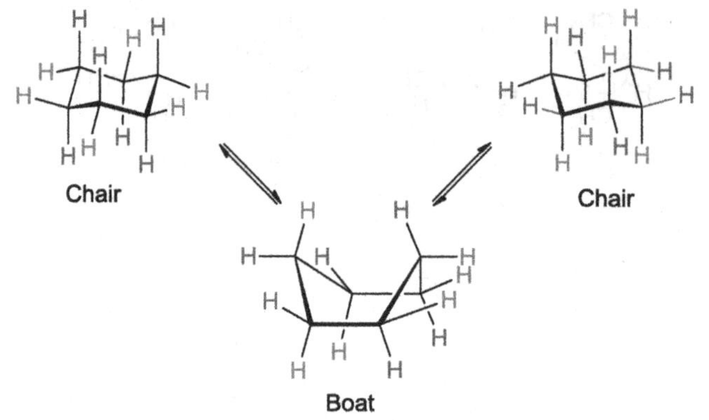

FIGURE 3.6
Conformational interconversions in cyclohexane.

cyclohexane, these two sets of hydrogens easily exchange between the two alternate chair conformers. However, this is not true when the ring has substituents that are not just hydrogen.

3.4 GEOMETRIC (*CIS-TRANS*) ISOMERS

Geometric isomers can be caused if a structural reason stops the bond rotation which would let the isomers convert from one to the other. This is common in both alkenes, because of the C=C, and cycloalkanes, because of the ring. Both of these features can cause substituents to have two different geometric relationships with one another. These relationships are called *cis*, which means on the same side, or *trans*, which means on the opposite side.

As mentioned earlier, we see that the structural isomers have different orders of their bonded atoms. However, geometric isomers have the same order of atomic bonding, but different arrangements in space. Any compounds that have the same order of atomic bonding, but different arrangement in space are called **stereoisomers**. Therefore, *cis-trans* isomers are one type of stereoisomers.

3.4.1 *Cis-Trans* Isomers in Cycloalkanes

With their cyclic structure, cycloalkanes can be looked at from two sides, a "top" side and a "bottom" side. Because of this, in substituted cycloalkanes, isomers become possible when the substituents have different geometric relationships to one another (Figure 3.7).

Cis-trans isomers are different compounds with different properties. To convert between these, strong covalent single σ bonds need to be broken and formed.

3.4.2 *Cis-Trans* Isomers in Alkenes

In compounds with a C=C, the bonding between the two sp^2-hybridized carbon atoms has a strong σ bond, ~340 kJ/mol, and a weaker π bond, ~270 kJ/mol.

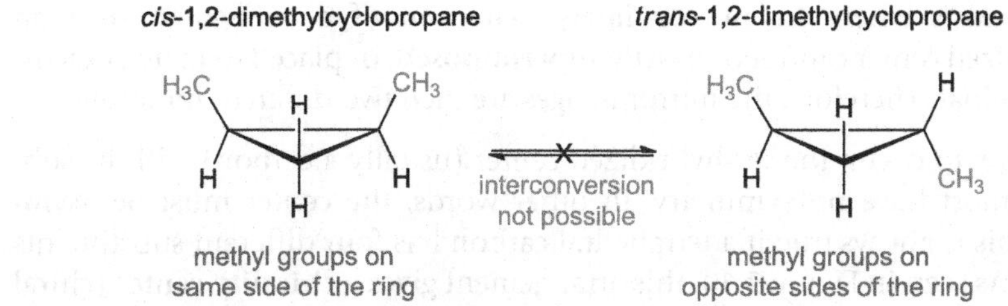

cis-1,2-dimethylcyclopropane   trans-1,2-dimethylcyclopropane

interconversion
not possible

methyl groups on
same side of the ring

methyl groups on
opposite sides of the ring

FIGURE 3.7
The *cis-trans* isomers of 1,2-dimethylcyclopropane.

The π bond is a weaker bond, but under normal conditions (about 60–80 kJ/mol), the bond is strong enough to stop rotation around the bond axis. As a result, when different substituents are attached to these carbon atoms, *cis* and *trans* isomers are possible. See Figure 3.8.

trans-2-butene   *cis*-2-butene

FIGURE 3.8
The *cis-trans* stereoisomers of 2-butene.

The stereoisomers in Figure 3.8 have different stabilities and properties. Under normal conditions, there is not enough energy for interconversion between this pair of geometric isomers. However, interconversion is relatively easier, using heat, catalysis, or photochemical action, because of the lower energy of the π bond. Note that if the two substituents at one end of a double bond are the same, then stereoisomers are not possible. See Figure 3.9.

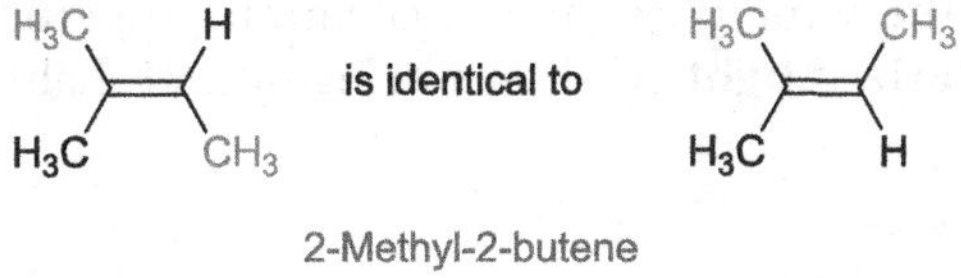

is identical to

2-Methyl-2-butene

FIGURE 3.9
Symmetry about the alkene double bond.

The situation is more complex when the alkenes are tri- and tetra-substituted. The *cis/trans* naming can no longer be used. Instead, a system of **sequence rules** (*E/Z* system) is used in a specific order to give the name. This system is not needed at this level, but details can be found in Appendix 5.

3.5 CONFIGURATIONAL ISOMERS

The word **configuration** describes the exact 3-D arrangement about an sp^3 tetrahedral center. This important type of stereoisomer occurs because carbon sp^3 tetrahedral centers can allow a molecule to show **chirality**.

Chirality exists when the **mirror image** reflection of a configuration at an sp^3-hybridized center cannot be exactly **superimposed**, or placed to match exactly, on the original. Therefore, the mirror images are then two different molecules.

For chirality to occur, the sp^3-hybridized center (usually a carbon) with its substituents must have no symmetry. In other words, the center must be **asymmetric**. This is always true if a tetrahedral carbon has four different substituents attached. As seen in Figure 3.10, this arrangement gives a **chirality center** (chiral center) and the molecule will be chiral. If two of the substituents are the same, then no chirality center exists and the molecule will be **achiral**.

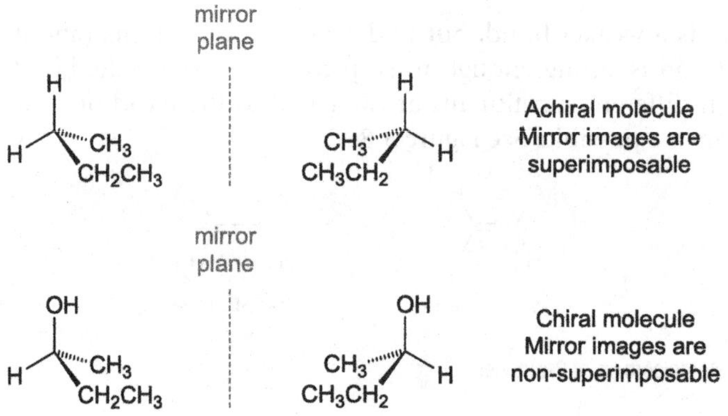

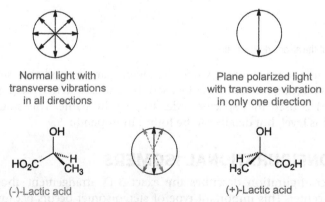

FIGURE 3.10
The origin of the carbon chirality center.

Chirality is a property of the whole molecule, but the cause of chirality is the chirality center within the molecule. **Enantiomers**, or an enantiomeric pair, are mirror image isomers that are not superimposable.

Enantiomers have the same physical and chemical properties except for their effect on **plane-polarized light**. As shown in Figure 3.11, light that has its waves

FIGURE 3.11
Plane-polarized light, the polarimeter, and molecular rotation.

filtered into a single plane is called plane-polarized light. Plane-polarized light is used in a **polarimeter** to measure the optical properties of enantiomers. Because of this, enantiomers are often called optical isomers.

When each enantiomer is placed in a polarimeter, the plane-polarized light is rotated in opposite directions, but in equal amounts. One enantiomer rotates the plane of light to the right (+), and the other rotates the plane to the left (–). An equal mixture of a pair of enantiomers is called a **racemic mixture**. If a racemic mixture is put in a polarimeter, the enantiomers cancel each other and the rotation is zero.

A chirality center is not the only possible cause of non-superimposable mirror images. However, for the purpose of this book, discussion is limited to chirality centers. The number of stereoisomers possible for any molecule can be calculated from the expression 2^n, where n is the number of chirality centers in the molecule. Examples of these more complex situations are shown in Chapter 8.

The (+) or (–) rotation of plane-polarized light by an enantiomer does not show the absolute configuration around the chirality center. To describe this absolute configuration, a set of sequence rules (R/S system) is used. This is outlined in Appendix 6.

3.6 SUMMARY OF ISOMER RELATIONSHIPS

Figure 3.12 summarizes the material in this chapter and shows the relationships between the various types of isomers.

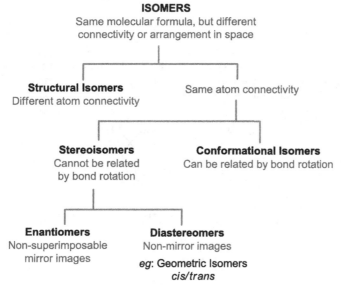

FIGURE 3.12
Isomer relationships.

QUESTIONS AND PROGRAMS

PROGRAM 7 Isomers

A The molecular formula of an organic compound can give lots of information about possible structural features. There are often many possible isomeric ways that the atoms can be joined together.

For example, because of tetravalency requirements, alkanes have molecular formulae that fit the general expression C_nH_{2n+2}. Alkenes fit the general formula C_nH_{2n}. Alkynes fit the general formula C_nH_{2n-2}.

✍ Draw a C_5 example of each to show that they fit the above formulae.

B Some simple examples that you might have drawn are:

$$CH_3CH_2CH_2CH_2CH_3 \qquad CH_3CH{=}CHCH_2CH_3 \qquad CH_3C{\equiv}CCH_2CH_3$$

Pentane C_5H_{12} 2-Pentene C_5H_{10} 2-Pentyne C_5H_8

Other examples are branched structural isomers. With more practice, you will see that the general formulae for alkanes, alkenes, and alkynes are equal to the loss of H_2 to increase the bond order from $C{-}C < C{=}C < C{\equiv}C$.

✍ What additional structures might also equal C_nH_{2n} and C_nH_{2n-2}?

C These general formulae equal the loss of H_2 and the formation of extra C–C bonds. Many different structures are possible. Instead of forming a $C{=}C$, the loss of H_2 can also give a cyclic structure. Both of these options give an additional C–C bond. For example, a formula of C_nH_{2n} can be for an alkyne, a diene, a cyclic alkene, or a bicyclic.

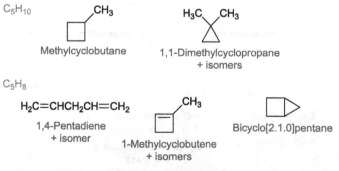

C_5H_{10}

Methylcyclobutane

1,1-Dimethylcyclopropane
+ isomers

C_5H_8

$H_2C{=}CHCH_2CH{=}CH_2$

1,4-Pentadiene
+ isomer

1-Methylcyclobutene
+ isomers

Bicyclo[2.1.0]pentane

✍ Is there an easy way to explain these observations?

D   One way to explain these facts is to use the formula equivalence between unsaturated and cyclic structures. It is very useful to know how many of these **double bond equivalents** (DBE) are present in a molecular formula. So far, in terms of the general formulae for hydrocarbons, this has been simple. What happens when the formula has heteroatoms in it?

The following simple equation gives the number of multiple bonds and/or rings in a molecular formula. This can be used for all neutral compounds that have carbon, hydrogen, nitrogen, oxygen, or halogens.

$$DBE = \frac{1}{2}\left[2n^4 + n^3 - n^1 + 2\right]$$

$$n^4 = \text{tetravalent atoms (carbons)}$$

$$n^3 = \text{trivalent atoms (nitrogens)}$$

$$n^1 = \text{monovalent atoms (hydrogens, halogens)}$$

✎ Try this equation on the formula $C_{10}H_{16}NO_3Br$.

E   Your calculation should have given a DBE=3. This answer shows that the molecule must have one of the following: 3 π bonds; 2 π bonds and 1 ring; 1 π bond and 2 rings; 3 rings.

The DBE might not seem very useful by itself, but it is usually used with other structural information such as functional groups. Whenever you suggest a structure for any molecule, it is useful to check if the DBE fits your structure.

Q 3.1. Give named structures that meet the following descriptions:
   (a) Five structural isomers with formula C_6H_{14}.
   (b) Four structural isomers with formula C_4H_8.
   (c) Six structural isomers of the alcohols with formula C_4H_8O.
   (d) Four structural isomers of the aldehydes with formula C_4H_6O.
   (e) Four structural isomers of the 1° or 2° amides with formula C_4H_9NO.

Q 3.2. Draw sawhorse diagrams of the two different staggered and two different eclipsed conformers of butane that are caused by rotation around the C_2–C_3 bond.

Q 3.3. Which of the conformational isomers in **Q 3.2** is the most stable (lowest energy). Why?

Q 3.4. Which of the following alkenes can have *cis/trans* stereoisomerism? Where *cis/trans* isomers are possible, draw named structures for the isomers.

(a) H_2C=$CCH_2CH_2CH_3$ with CH_3

(b) CH_3C=$CHCH_2CH_3$ with CH_3

(c) CH_3CHCH=$CHCH_3$ with CH_3

(d) CH_3CHCH_2CH=CH_2 with CH_3

Q 3.5. Which of the following molecules can show *cis/trans* stereoisomerism? Where *cis/trans* isomers are possible, draw a structure for the *cis*-isomer.

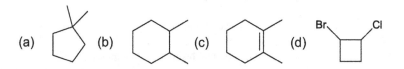

(a)   (b)   (c)   (d)

Q 3.6. Draw structures for the following compounds and use these to decide if *cis/trans* stereoisomerism is possible. If it is, draw the *trans*-isomer.

(a) 2-Methyl-3-hexene
(b) 2-Methyl-2-hexene
(c) 2-Methyl-1-butene
(d) 3-Ethyl-3-methyl-1-pentene
(e) Vinylcyclopropane
(f) 1-Isopropyl-4-methylcyclohexene
(g) 2,5-Dimethyl-2-hexene
(h) 1,4-Dichlorocyclohexane

Q 3.7. Draw the compound that fits the information provided in the following text.

(a) Molecular formula C_6H_{10}, three methyl groups.
(b) C_5H_{10}, no *cis-trans* stereoisomerism, and has one ethyl group.
(c) $C_3H_6O_2$, has an aldehyde and a secondary alcohol.

Q 3.8. For the compound provided:

$$O\text{---}CH_2CH_3$$
$$H_3C\diagdown \; H \diagup CH_2$$
$$C$$
$$HC{\equiv}C\diagup \overset{|}{\underset{\|}{C}}\diagdown C \diagup \overset{H}{\underset{H}{C}}{=}\overset{}{C}\diagdown CH_3$$
$$O$$

Indicate:

(a) A part of the molecule with three carbon atoms that lie in a straight line.
(b) A tertiary carbon atom.
(c) An sp^2-hybridized carbon.
(d) An ethoxy group.
(e) A part of the molecule that can have geometric isomerism.

PROGRAM 8 Chirality

A You need to be able to quickly identify molecules that are chiral because they have a chirality center. This needs you to practice with many examples. The best way to find a chirality center is to study the tetrahedral carbon centers and their substituents.

$$CH_3CH_2CH_2CH\,(OH)\,CH_2CH_3$$

🐌 Is the 3-hexanol chiral?

B It is easier if you draw the structure in more detail. Then it is easy to see that the CH_2 and CH_3 groups cannot be chirality centers because they do not have four different attachments.

The remaining carbon **C** has H, OH, CH_2CH_3, and $CH_2CH_2CH_3$ as the four different ligands. Note that the ethyl and propyl substituents are each taken as a complete unit. Therefore, the molecule is chiral.

$$CH_3CH_2CH_2\!-\!\overset{\displaystyle OH}{\underset{\displaystyle H}{C}}\!-\!CH_2CH_3$$

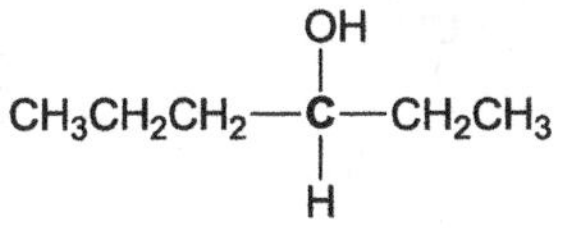

🐌 Is 2-methylcyclohexanone chiral?

C You should have drawn a structural diagram that clearly shows the attachments at each of the tetrahedral centers.

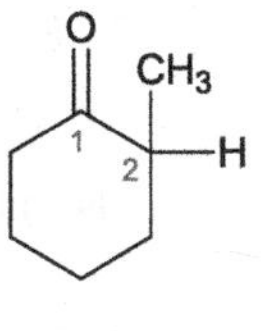

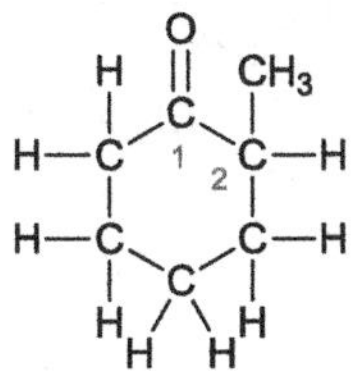

The CH_3 group, the four ring CH_2, and the C=O cannot be chirality centers. However, the ring C_2 is bonded to four different substituents, the CH_3, H, C=O, and CH_2, and is a chirality center. Therefore, the molecule is chiral.

🐌 What happens if the above is the structural isomer 3-methylcyclohexanone?

D This example shows that sometimes you need to look further than the first point of attachment. You saw this in **B** where the ethyl and propyl ligands were identified as different groups.

This is easy to see in acyclic examples, but it is often more difficult to see in cyclic examples.

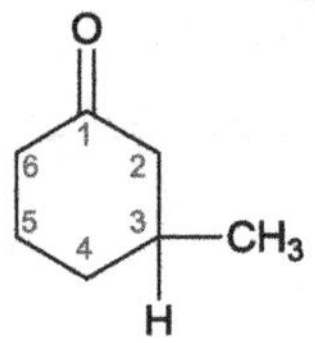

Here you only need to look at the C_3 center. The CH_3 and H ligands are simple to see. The other two substituents, the CH_2 groups at C_2 and C_4, seem equal at first. As you move around the ring in either direction, these groups become different at C_1 and C_5. Therefore C_3 is a chirality center and the molecule is chiral.

Q 3.9. Mark all of the chirality centers in each of the following molecules.

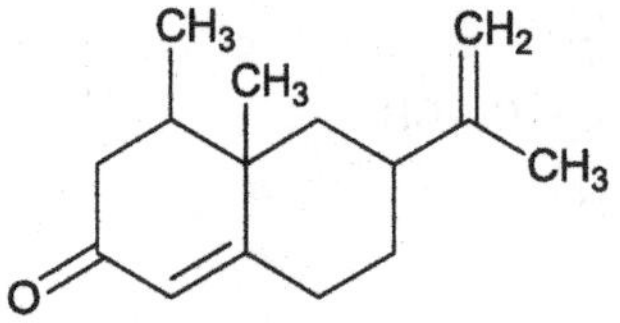

Nootkatone
(grapefruit oil)

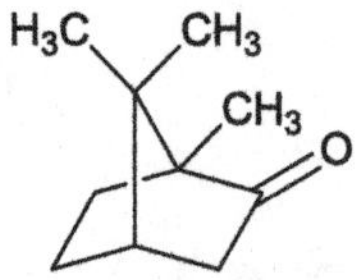

Camphor

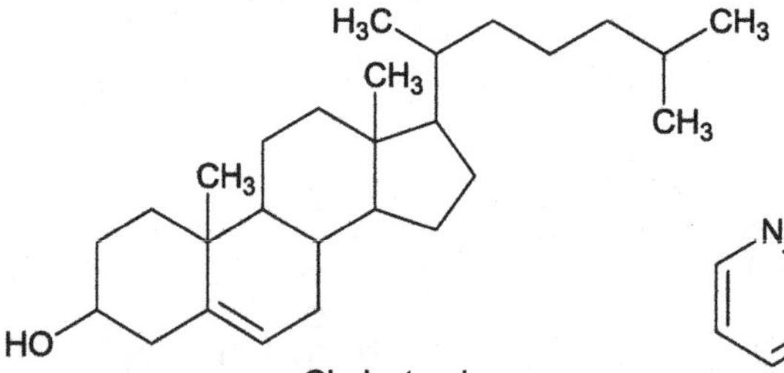

Cholesterol

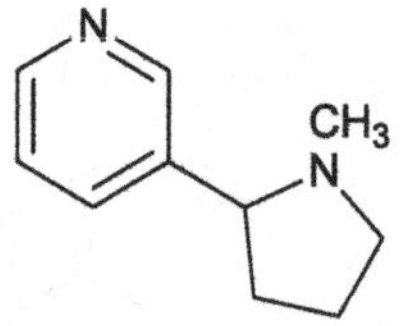

Nicotine

CHAPTER 4

Resonance and Delocalization

4.1   WHAT IS RESONANCE?

In a molecular structure, there is often more than one way to draw the π-bonding and/or non-bonding lone pair of electrons, without changing the atom connections. These are examples where a single Lewis structure does not give a complete picture of the bonding in the molecule. The concept of **resonance** can be used to describe the bonding in these situations.

As Figure 4.1 shows, the bonding in a carbonate dianion is described by a **resonance hybrid** (an average) of three equal **resonance forms** (alternative Lewis structures). Rather than have the bonding electrons localized, the resonance hybrid has the available electrons delocalized over the bonded atoms.

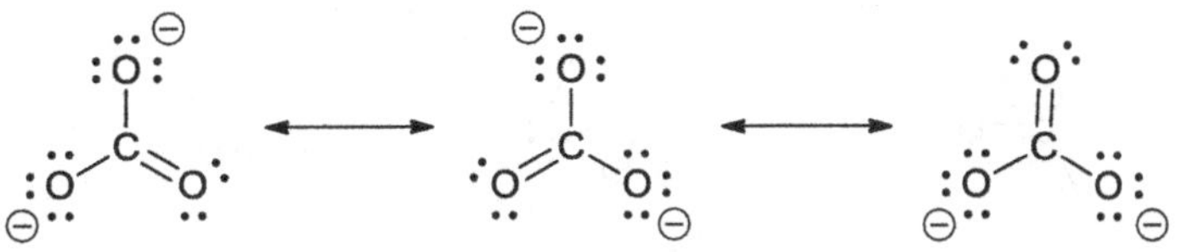

FIGURE 4.1
Resonance forms for the CO_3^{2-} carbonate dianion.

The three oxygen atoms share the π and/or non-bonding lone pair of electrons and the negative charges equally. This resonance model represents accurately the experimental measurement of the three C–O bonds in the dianion. All C–O bonds are:

■ the same in length;
■ a little shorter than a standard C–O single bond;
■ a little longer than a standard C=O double bond.

4.2   DRAWING USEFUL RESONANCE STRUCTURES

The following simple guidelines will help you to draw and understand resonance structures:

■ Individual resonance forms are not real. The only real structure is a hybrid of the resonance forms. As shown in Figure 4.1, a double-headed arrow ↔ is used to link resonance forms. This type of arrow shows only a resonance situation and does not show an equilibrium between real structures.

Organic Chemistry Concepts: An EFL Approach. http://dx.doi.org/10.1016/B978-0-12-801699-2.00004-3

- Resonance forms differ from each other only in the arrangement of the π and/or non-bonding lone pairs of electrons. The octet rule of valency is followed. The relative position of all bonded atoms remains unchanged, as in Figure 4.1. For comparison, Figure 4.2 shows a non-resonance situation.

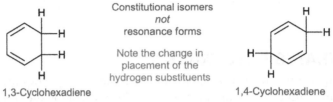

Constitutional isomers
not
resonance forms

Note the change in
placement of the
hydrogen substituents

1,3-Cyclohexadiene   1,4-Cyclohexadiene

FIGURE 4.2
The non-resonance relationship between dienes.

- Resonance forms do not have to be equal. As Figure 4.3 shows, each Lewis structure contributes to the overall resonance hybrid, but these contributions are not always equal. If resonance forms are not equal, the resonance hybrid will be more like the Lewis structure of the most stable resonance form. The hybrid is a weighted average of the resonance forms.

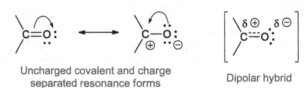

Uncharged covalent and charge
separated resonance forms

Dipolar hybrid

FIGURE 4.3
Resonance description of the carbonyl group.

- A resonance situation is more stable than a non-resonance situation. The resonance hybrid is more stable than any single resonance form. The more reasonable resonance forms that can be drawn, the greater the stability of the molecule.

In the below benzene system, as shown in Figure 4.4,

- all six C–C bonds are the same length, between average single and double bond lengths;
- the potential energy of the delocalized hybrid is 152 kJ/mol lower than the predicted potential energy of either resonance form with a fixed bonding arrangement.

Resonance forms of benzene
showing localized π–bonding
electron arrangements

Resonance hybrid
showing delocalized
bonding arrangement

FIGURE 4.4
Resonance forms for benzene.

4.3 USING CURLY ARROWS TO COUNT ELECTRONS

Curly arrows are a convenient way to show electron movement. As shown in Figures 4.3 and 4.4, they also help you to count electrons. Chapter 5 uses curly arrows to show the details of the bond breaking and formation in reaction mechanisms.

The following rules apply:

■ A full-headed arrow ⟳ shows the movement of an electron pair;
■ A half-headed arrow ⟳ shows the movement of a single electron;
■ All arrows start at the point on the structure that is giving the electron(s). This point could be a formal charge, a π-bond, or a lone pair;
■ The arrow head then points exactly to the place on the structure to where the electron movement is going;
■ The direction of the arrow depends on the atom types and bonding involved.

For example, Figure 4.5 shows that it is possible to redraw Figure 4.3 with the curly arrow showing the opposite direction of electron movement. However, because oxygen is more electronegative than carbon, this is less likely. The charge-separated resonance form in Figure 4.5 has a very high energy. Therefore, it makes only a very small contribution to the resonance hybrid and is usually ignored.

Less likely charge
separated resonance form

FIGURE 4.5
Alternate resonance forms for the carbonyl group.

QUESTIONS AND PROGRAMS

PROGRAM 9 Drawing Resonance Forms

A   Drawing reasonable resonance forms for molecular structures helps us understand the stability and reactivity of molecules.

Using curly arrows helps with this by making sure that no electrons are lost or gained. Some guidelines are:

(1) All resonance forms have the same geometric arrangement of atoms.
(2) No single resonance form should have unlikely high energy. This means that:
   (a) All resonance forms should meet the octet rule for atoms that have atomic number 10 or less (C, H, O, N, F).
   (b) All resonance forms should have an equal number of electron pairs.
(3) All resonance forms have the same overall charge.
(4) No single resonance form gives a true picture of the molecule. Their weighted average together makes the resonance hybrid.

For example, reasonable resonance forms for the propanone molecule are:

The curly arrow shows movement of the π-bonded electron pair onto the electronegative oxygen. This is the same as drawing that bond in the ionic form. Note the use of the double-headed arrow between resonance forms.

🐭 Try to draw other resonance forms for propanone.

B   Other electron movements that you could have drawn for the π-bond are:

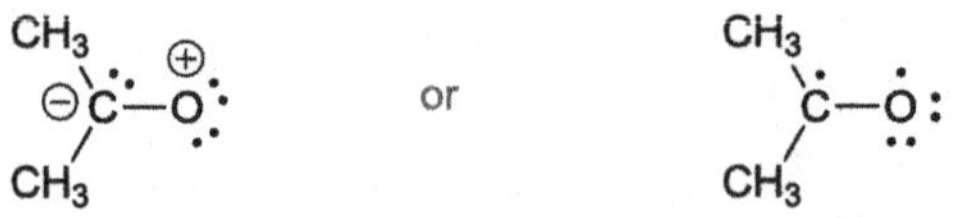

These are both relatively high energy resonance forms. The first form shows the unlikely movement of the π-bonded electron pair away from the more electronegative oxygen toward the carbon. The second form shows the equal splitting of the π-electron pair to give radicals.

How much a resonance form affects the resonance hybrid depends on its relative energy. Because these resonance forms have much higher energies, they have little effect on the final resonance hybrid.

🐭 Now draw the resonance hybrid for propanone.

C Your structure must show the effects from each of the two most likely resonance forms. For example,

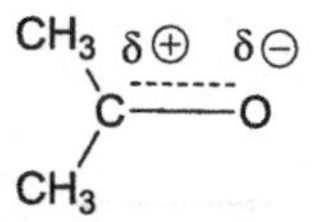

$$\begin{array}{c} CH_3 \\ \diagdown \\ \diagup C \text{------} O \\ CH_3 \end{array} \quad \delta \oplus \quad \delta \ominus$$

This hybrid shows the polar nature of the C=O which is bond caused by the high electronegativity of the oxygen atom. The structure shows the effect from the resonance form with both a positive and a negative charge.

We usually draw the carbonyl group as C=O, but it is important to remember these resonance forms. This concept is important later when you study reactivity and stability for the C=O group.

Q 4.1. Draw the resonance structure that is formed from the curly arrow movements shown on the following.

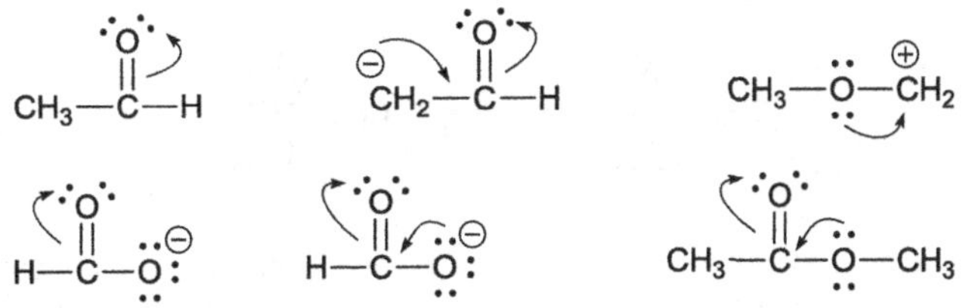

$$CH_3-\overset{\overset{\displaystyle \cdot\overset{..}{O}\cdot}{\|}}{C}-H \qquad \overset{\ominus}{CH_2}-\overset{\overset{\displaystyle \cdot\overset{..}{O}\cdot}{\|}}{C}-H \qquad CH_3-\overset{..}{\underset{..}{O}}-\overset{\oplus}{CH_2}$$

$$H-\overset{\overset{\displaystyle \cdot\overset{..}{O}\cdot}{\|}}{C}-\overset{..}{\underset{..}{O}}: \ominus \qquad H-\overset{\overset{\displaystyle \cdot\overset{..}{O}\cdot}{\|}}{C}-\overset{..}{\underset{..}{O}}: \ominus \qquad CH_3-\overset{\overset{\displaystyle \cdot\overset{..}{O}\cdot}{\|}}{C}-\overset{..}{\underset{..}{O}}-CH_3$$

PROGRAM 10 Evaluation of Resonance Forms

A The resonance hybrid is always more stable than any single resonance form. Symmetrical (equal) resonance forms are the most effective and produce greater energy lowering by resonance.

It is important to evaluate the relative effect of a resonance form on the hybrid. We can ignore very high energy resonance forms, and evaluate the efficiency of the resonance as shown in Program 9B.

✎ Draw and evaluate the possible resonance forms for ethylene.

B One possible set of resonance forms is:

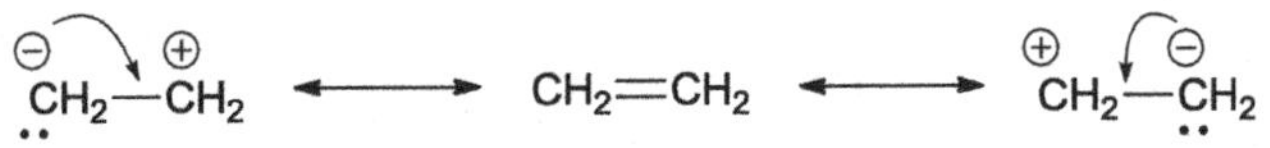

$$\overset{\ominus}{\underset{..}{CH_2}}-\overset{\oplus}{CH_2} \quad \longleftrightarrow \quad CH_2=CH_2 \quad \longleftrightarrow \quad \overset{\oplus}{CH_2}-\overset{\ominus}{\underset{..}{CH_2}}$$

Because carbon prefers not to carry any charge, the charge-separated resonance forms have too high an energy to be practical. Unlike oxygen, carbon is less able to carry a negative charge, and the positive charge leaves carbon with an incomplete octet.

Continued...

—Cont'd

Therefore, the effect of these charge-separated resonance forms on the overall hybrid is very small. Compare this to Program 9A in which the propanone situation is very different.

Here, the electronegative oxygen makes the dipolar resonance form more acceptable. However, the dipolar resonance form, in which carbon has an incomplete octet, has a higher energy and contributes less to the hybrid.

✎ Draw a system that shows equal resonance forms.

C One possible example is the nitromethane molecule.

Because these two resonance forms are equal in structure, they have the same energy. Experimental results show that both N–O bonds are equal in length. They are shorter than a single N–O bond, but they are longer than a double N–O bond. In this case, the resonance hybrid is a simple average of the two resonance forms.

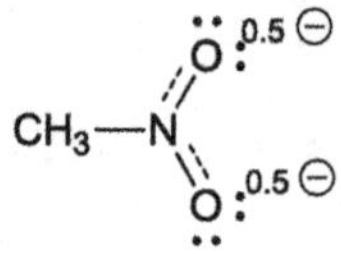

Note that curly arrows are used to switch between resonance forms by moving pairs of electrons between atoms and adjacent bonds. The atoms that receive or give these pairs only gain a single charge. The reason is that these atoms only get or lose a share in one of the electrons of the pair. Go back to the examples mentioned earlier and check this for yourself.

In general structure drawing, it is common to leave out the lone pairs. Do not forget that they are still there and that they must be included to help with electron counting.

Q 4.2. Study the pairs of resonance structures in Q 4.1 and try to evaluate their relative contribution to the overall hybrid.

PROGRAM 11 Resonance in Conjugated Systems

A Programs 9 and 10 show how electron pairs are moved from atoms to adjacent bonds. This program shows that it is possible to move these pairs from bond to adjacent bond. Study the positively charged species below.

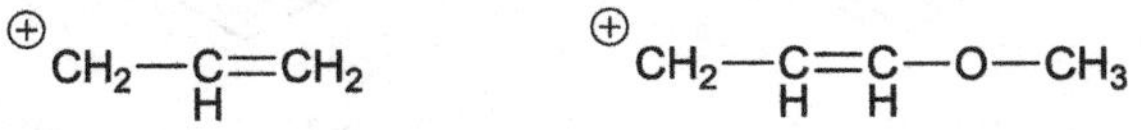

🖎 Write resonance forms for these carbocations.

B You should have drawn the following.

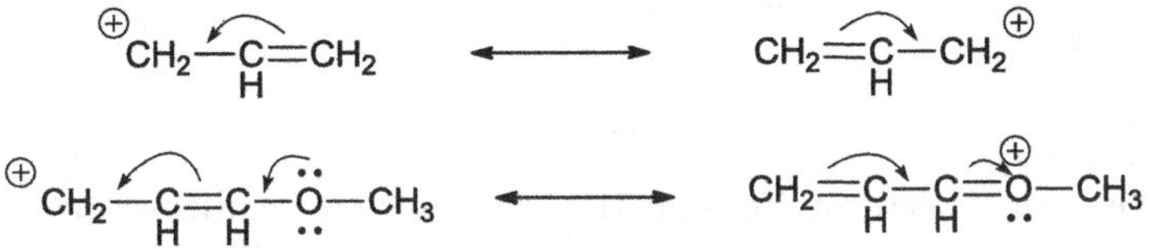

The first example is the allylic system. It shows two equal resonance forms. The second example gives two unequal structures. Note how the lone pairs on the oxygen are used in the curly arrow electron movements.

🖎 Now use the same method to draw resonance forms for the uncharged systems below.

C The two most important resonance forms for each are:

The dipolar form in the first example is the minor one, but it is still important. It is a simple extension of the concept shown earlier in Program 9A. It shows the electron movement extended into the adjacent conjugated C=C double bond. This structure helps explain the effect that the electronegative oxygen has on the carbon center that is three atoms away.

The second example simply shows the two localized structures that are the major resonance forms for the aromatic benzene system.

Q 4.3. Use the ideas from Program 11 to draw resonance forms for:

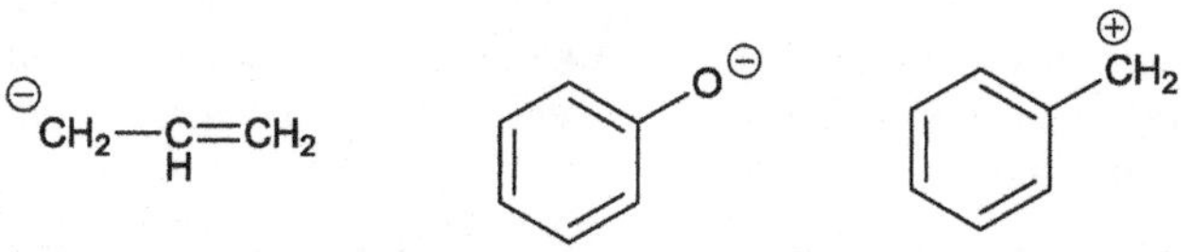

PROGRAM 12 Delocalization

A This book does not give a full description of the resonance concept in aromatic systems. The Programs mentioned earlier are the basis to understand delocalization, especially of charged systems. This helps us understand the relative reactivity and stability of species.

The concepts explained earlier only show ionic species. This can also be done for radical species with unpaired electrons by using curly arrows with single heads. A single-headed arrow shows the movement of single electrons rather than pairs. For example, resonance delocalization in the allylic radical is shown below.

B In the examples mentioned earlier, we can see the important role of conjugation. Conjugation can occur between any species that are separated by no more than one single bond. If a system is not conjugated, then delocalization is not possible. The following example shows this principle.

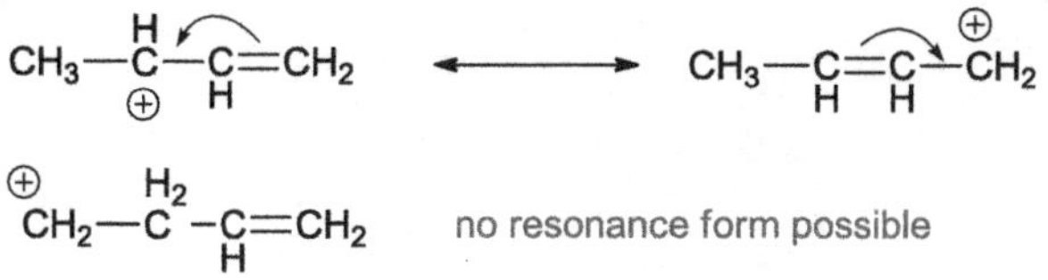

$$CH_3-\overset{H}{\underset{\oplus}{C}}-\overset{}{\underset{H}{C}}{=}CH_2 \longleftrightarrow CH_3-\overset{}{\underset{H}{C}}{=}\overset{}{\underset{H}{C}}-\overset{\oplus}{C}H_2$$

$$\overset{\oplus}{C}H_2-\overset{H_2}{C}-\overset{}{\underset{H}{C}}{=}CH_2 \qquad \text{no resonance form possible}$$

Delocalization is possible for the first cation. However, no similar resonance form can be drawn for the second cation because two single bonds lie between the charged carbon and the C=C. This is important because the conjugated system is more stable than the non-conjugated one.

🐭 Use the above idea on the uncharged systems below. What can you say about their relative stability?

$$CH_3-\overset{}{\underset{H}{C}}{=}\overset{}{\underset{H}{C}}-\overset{}{\underset{CH_3}{C}}{=}O \qquad\qquad CH_2{=}\overset{}{\underset{H}{C}}-\overset{H_2}{\underset{CH_3}{C}}-C{=}O$$

C The first of these systems is conjugated because it has alternating double and single bonds. Resonance forms can be drawn as shown below. In the second system, the double bonds are isolated from one another, and no similar resonance forms can be drawn.

$$CH_3-\overset{}{\underset{H}{C}}{=}\overset{}{\underset{H}{C}}-\overset{}{\underset{CH_3}{C}}{=}O \longleftrightarrow CH_3-\overset{H}{\underset{\oplus}{C}}-\overset{}{\underset{H}{C}}{=}\overset{}{\underset{CH_3}{C}}-\overset{\ominus}{O}$$

The conjugated system is more stable and has lower energy than the non-conjugated one. The aromatic system in Program 11 is a perfect example of the efficient use of conjugation.

🐭 What is the possible situation with the following molecules?

ketene $CH_2{=}C{=}O$ **allene** $CH_2{=}C{=}CH_2$

D Because ketene and allene have no single bonds between the double bonds, they are not conjugated systems. These cumulated double bonds may at first seem good for electron flow between adjacent bonds. However, you will find that it is not possible to draw resonance forms for these.

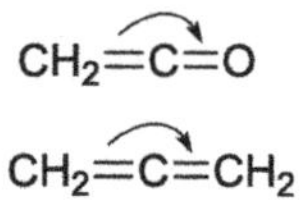

$CH_2{=}C{=}O$ If you move a pair of electrons, either oxygen or carbon gets an expanded octet - a forbidden

$CH_2{=}C{=}CH_2$ operation

CHAPTER 5
Reactivity: How and Why

5.1 WHY DO REACTIONS OCCUR?

This chapter will show you how to use the concepts of shape and electronic structure of organic compounds to find out what types of reaction the compounds undergo. We use the models of hybridization, inductive effect, and delocalization to help us understand how and why organic reactions occur.

5.2 BOND BREAKING AND MAKING

Bonds are broken and made in all chemical reactions. With covalently bonded atoms, there are two ways for the two-electron bond to break. These are shown in Figure 5.1.

The first way, **homolysis**, is the symmetrical sharing of the bonding electrons and leads to two radicals. Each **radical** has one unpaired electron from the original bonding pair. The second way, **heterolysis**, is the unsymmetrical sharing of the bonding electrons and leads to ions. The negatively charged anion has both original bonding electrons and the positively charged cation has an empty orbital.

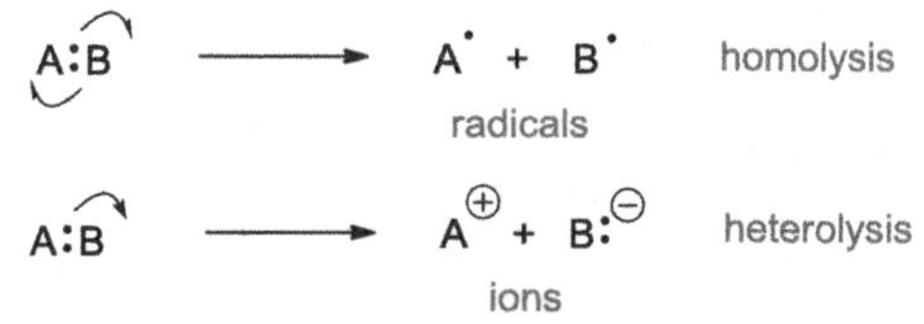

FIGURE 5.1
Bond-breaking processes.

Figure 5.2 shows that the reverse is true for covalent bond making. If radicals combine, we call this **homogenic** bond making. If a cation and an anion combine, we call this **heterogenic** bond making.

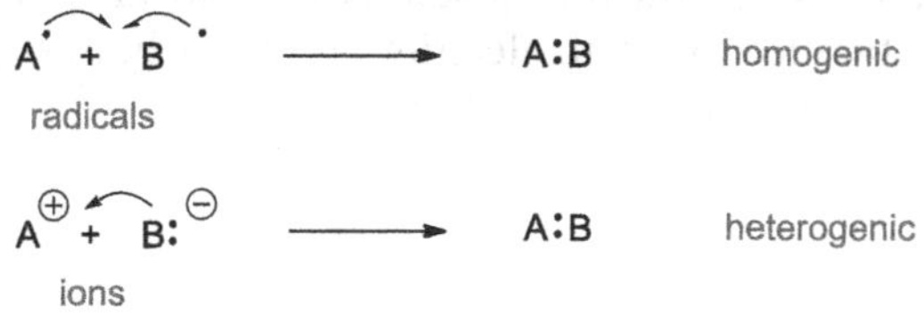

FIGURE 5.2
Bond-making processes.

Organic Chemistry Concepts: An EFL Approach. http://dx.doi.org/10.1016/B978-0-12-801699-2.00005-5

5.3   REACTIVE SPECIES

The breaking and making of bonds in Figures 5.1 and 5.2 show the formation and use of reactive species such as radicals and ions. These processes usually use diagrams with curly arrows to help with accurate electron counting.

Bonds to carbon are the focus of most organic reactions. Figure 5.3 shows a curly arrow description of how carbon reactive species are formed. Homolysis gives radicals, and heterolysis gives ions. These ions of carbon can either be a **carbanion** or a **carbocation**. A carbanion is a negatively charged carbon, and a carbocation is a positively charged carbon.

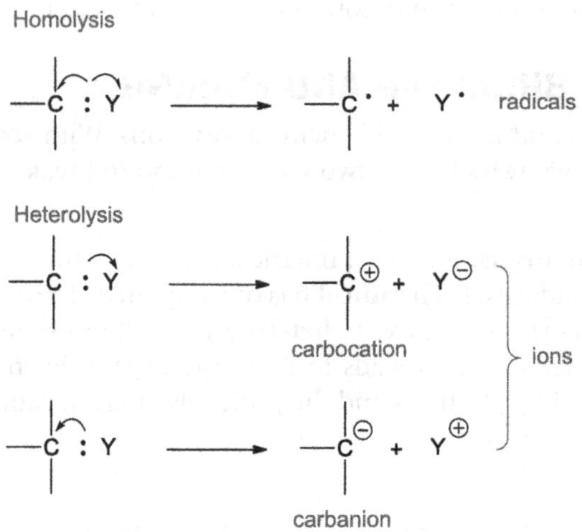

FIGURE 5.3
Formation of carbon reactive species.

Figure 5.3 shows reactive species that are easy to find. This is because the species carry an ionic charge or radical. The concept of reactivity goes further than these obvious reactive species. It is more useful to generally classify reactive species as **electrophiles** or **nucleophiles**. Electrophiles are any species that are electron poor. Nucleophiles are any species that are electron rich.

This classification includes more than just ions and radicals. It also includes species that have non-bonded lone pair electrons, multiple bonds, and highly polar bond dipoles. See Figure 5.4.

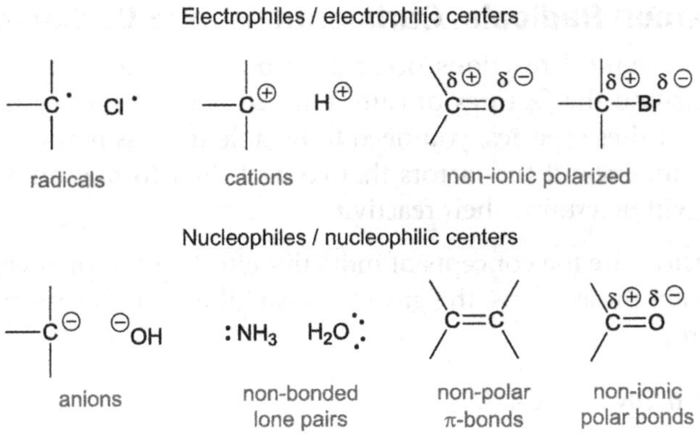

Electrophiles / electrophilic centers

radicals   cations   non-ionic polarized

Nucleophiles / nucleophilic centers

anions   non-bonded   non-polar   non-ionic
              lone pairs   π-bonds   polar bonds

FIGURE 5.4
Electrophile and nucleophile classification.

Any atomic center which is electron poor can act as an electrophile. Electron-poor centers are caused by:

- bond breaking to give radicals or cations;
- inductive effects as a result of electronegativity differences in a bond.

Any atomic center which is electron rich can be a nucleophile. Electron-rich centers can be caused by:

- bond breaking to give anions;
- lone pairs of electrons;
- non-polar multiple bonds;
- inductive effects as a result of electronegativity differences in a bond.

Therefore most organic reactions are the combination of electrophiles with nucleophiles. It is important to learn to find them. Two of the examples mentioned earlier need further explanation.

- When there are multiple bonds between the same atoms (usually carbon), the bond is largely non-polar. The area between the two bonded atoms has a relatively high electron concentration, so it is nucleophilic. This occurs for either double bonds with four electrons or triple bonds with six electrons.
- Heteroatoms such as nitrogen, oxygen, and halogens cause large electronegativity differences by inductive effects. This gives polar single or multiple bonds. This makes one atomic center relatively electrophilic and the other relatively nucleophilic.

5.3.1 Carbon Radicals, Carbocations, and Carbanions

Most organic chemical reactions occur at carbon. Therefore, it is very important to understand the features of carbon-based reactive species. To deal with the reactions of these species, you need to be able to classify them. To do this, you need to understand the factors that control their formation and stability, because this will determine their reactivity.

The major factors are the concepts of inductive effect and resonance/delocalization. The greater these effects, the greater the stability, and the easier it is for the species to form.

5.3.1.1 CARBON RADICALS

As seen in Figure 5.3, radicals are formed by homolytic bond breaking. With only three ligands, radicals try to have a planar sp^2 shape. Stability is determined by the ability of the attached ligands to feed electron density to the electron-poor radical center. Electron donation is by either inductive effects or resonance/delocalization.

- Inductive effects.

This gives a general stability order of $3° > 2° > 1°$.

- Resonance/delocalization.

5.3.1.2 CARBOCATIONS

As seen in Figure 5.3, carbocations are formed by heterolytic bond breaking. With only three ligands, they also try to have a planar sp^2 shape. Stability is determined by the ability of the attached ligands to feed electron density to the electron-poor carbocation. Electron donation is by either inductive effects or resonance/delocalization.

- Inductive effects.

This gives a general stability order of $3° > 2° > 1°$.

■ Resonance/delocalization.

$$CH_2=\overset{\underset{|}{H}}{C}-\overset{\oplus}{C}H_2 \quad \longleftrightarrow \quad \overset{\oplus}{C}H_2-\overset{\underset{|}{H}}{C}=CH_2$$

5.3.1.3 CARBANIONS

As seen in Figure 5.3, carbanions are also formed by heterolytic bond cleavage. With three ligands and an electron pair, they try to keep their tetrahedral sp^3 shape. Stability is determined by the ability of the attached ligands to share the excess electron density. This sharing is by either inductive effects or resonance/delocalization.

■ Inductive effects.

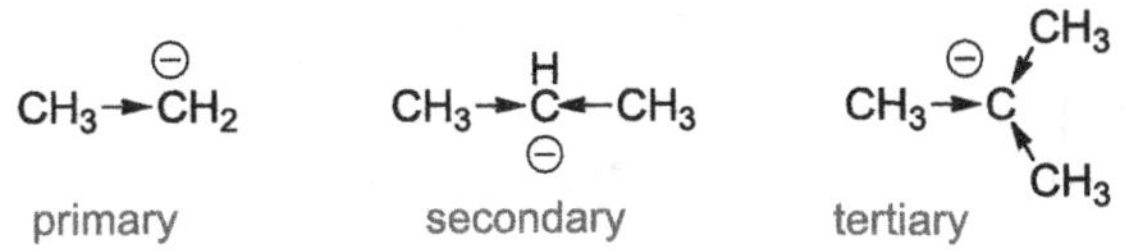

$$\overset{\ominus}{CH_3 \rightarrow CH_2} \qquad CH_3 \rightarrow \overset{\underset{\ominus}{\overset{H}{|}}}{C} \leftarrow CH_3 \qquad CH_3 \rightarrow \overset{\ominus}{C}\overset{CH_3}{\underset{CH_3}{<}}$$

primary   secondary   tertiary

This gives a general stability order of $1° > 2° > 3°$.

■ Resonance/delocalization.

$$CH_2=\overset{\underset{|}{H}}{C}-\overset{\ominus}{C}H_2 \quad \longleftrightarrow \quad \overset{\ominus}{C}H_2-\overset{\underset{|}{H}}{C}=CH_2$$

5.4 REACTION TYPES

At first, organic chemistry seems to have a very large number of different reactions that need to be memorized. It is true that many thousands of reaction examples exist, but there are only four main reaction types. These are:

■ **addition**, in which substituents are added to multiple bonds;
■ **elimination**, in which substituents are lost to give a multiple bond;
■ **substitution**, in which one substituent is replaced by another;
■ **rearrangement**, in which a substrate changes to a structural isomer.

All reaction examples fall into one of these four categories. Therefore, organic chemistry has many different versions of these four primary reaction types. These reaction types can be polar or ionic and occur with heterolytic bond processes, or they can be radical and occur with homolytic bond processes.

5.4.1 Addition Reactions

In an addition reaction, the reacting species are simply added together to give a single product. In most examples, a reagent ligand is added to either end of a multiple bond. This causes a decrease in the bond order and an increase in hybridization state. For example, Figure 5.5 shows an addition causing a change from sp^2 to sp^3.

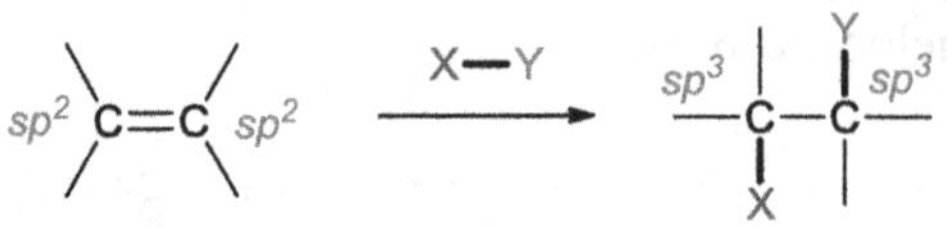

FIGURE 5.5
A typical addition reaction.

5.4.2   Elimination Reactions

An elimination reaction is the reverse of addition. In most examples, a ligand is removed from each of two adjacent bonded centers. This causes an increase in bond order and a decrease in hybridization state. For example, Figure 5.6 shows an elimination causing a change from sp^3 to sp^2.

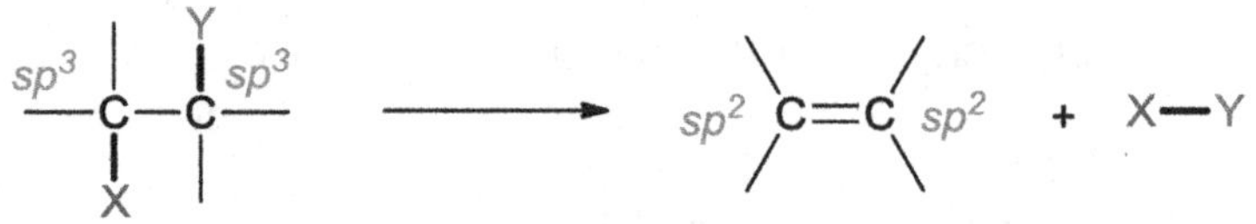

FIGURE 5.6
A typical elimination reaction.

5.4.3   Substitution Reactions

A substitution reaction is the exchange of ligands between two reactants. Bond order and hybridization remain the same in substitution. See Figure 5.7.

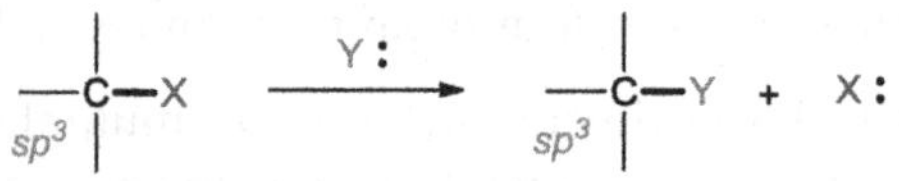

FIGURE 5.7
A typical substitution reaction.

5.4.4   Rearrangement Reactions

In the simplest examples, a rearrangement reaction is a reorganization of the bonds and atoms within a reactant to give a structural isomer. Figure 5.8 shows an example of *cis–trans* isomerism.

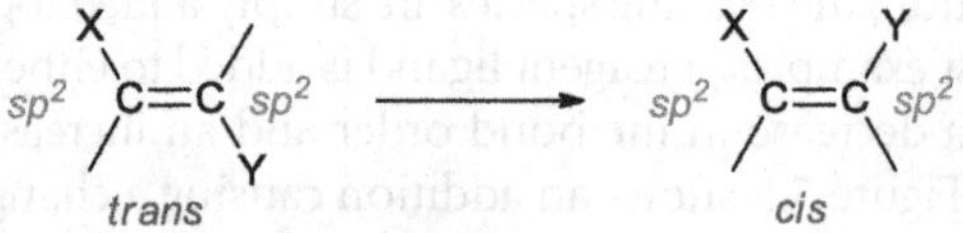

FIGURE 5.8
A typical rearrangement reaction.

In this introduction to concepts, we do not discuss rearrangements in any detail. It is enough to cover the range of addition, elimination, and substitution, because they represent the majority of organic reactions. For these three reaction types, the focus can be either on those involving polar or ionic reactive species. Examples with radicals are less common and are not needed for a general understanding of the basic principles.

5.5 REACTION MECHANISM: THE PATH FROM REACTANT TO PRODUCT

For an organic chemist, a **reaction mechanism** is a detailed description of the way bonds are broken and made as a reaction goes from starting materials to products. If they are not part of the changes, solvents or inorganic additives are usually ignored.

A mechanism is a proposal that shows electron movements which explain all bond breaking and making. For example, see Figure 5.9. The main information that a mechanism should show includes:

- any reactive intermediates or transition states that occur;
- curly arrows to show all electron movements;
- information about relative reaction rates in stepwise reactions;
- details of any relevant stereochemical features.

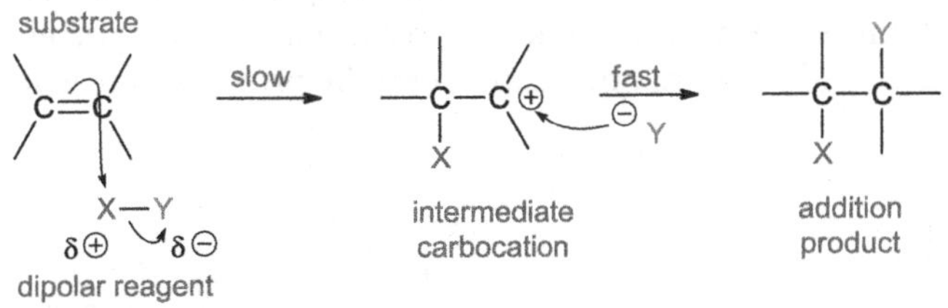

FIGURE 5.9
A mechanism for an addition reaction.

The difference between a **reactive intermediate** and a **transition state** can be explained as follows. A reactive intermediate is a real species in a reaction and it can be measured. This can help to prove a suggested mechanism.

A transition state is a reasonable proposal of a species that can occur in a reaction mechanism. However, this species cannot be proven by measurement. Transition states can be suggested at any point along the reaction path which a mechanism follows from one step to the next. Important transition states are often at the points of highest energy during the progress of a reaction. See examples in Section 5.6.

Curly arrows are an important tool to describe mechanisms. They show details of what happens to the nucleophiles, electrophiles, and radicals that are formed in the reaction. They allow us to follow all bond breaking and making, and help us to accurately count electrons. Remember that curly arrows are specific and are always drawn from the electron-rich source to the electron-poor center.

For our purposes, we will keep all data of the reaction rates as **qualitative**. Qualitative means that we do not need exact values. We only use relative terms, such as fast or slow, to describe the ease or rate of a reaction step. The slowest step in a stepwise reaction controls the overall rate of reaction. This is called the **rate-determining step**. Changes in stereochemistry during reactions can occur. Further details are shown in Appendices 8 and 9.

5.6 REACTION ENERGY

Qualitative reaction energetics uses the basic concepts of:

- **reaction rate**–how fast a reaction occurs;
- **reaction equilibrium**–shows the overall direction of a reaction.

For a reaction to be spontaneous and give off energy, it must have an equilibrium constant >1. Then the reaction is described as **exothermic**, and the heat of reaction ΔH is negative.

If energy must be added to make a reaction occur, the equilibrium constant is <1. Then the reaction is described as **endothermic**, and ΔH is positive. This qualitative description ignores the relatively small entropy contribution.

Energy diagrams, as shown in Figure 5.10, are simple qualitative descriptions of the energy changes during a reaction. The vertical axis shows the energy of the reacting system, and the horizontal axis shows the progress of reaction.

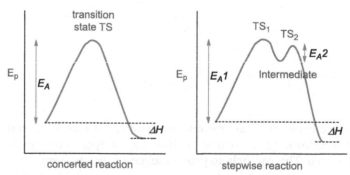

FIGURE 5.10
Typical reaction energy diagrams.

As Figure 5.10 shows, a **concerted**, single-step process has a single **activation energy** (E_A) that must be provided for reaction to occur. This leads to the energy maximum of the transition state (TS). In the example shown, ΔH is negative and means an overall exothermic process. The rate of the reaction depends on the size of E_A and the ability of the reacting molecules to overcome this energy barrier.

For a **stepwise** reaction, because there is more than one step, a true reactive intermediate exists. The steps on each side of the intermediate can be seen as two separate processes, each with their own activation energies (E_A1 and E_A2). These steps each have their own reaction rates depending on their activation energies. In this example, the overall rate-determining step has the larger E_A1. Because the ΔH is negative for the overall reaction, it is an exothermic example.

The reactive intermediates which are discussed in this book are mostly carbocations and carbanions. Note that these could also be radicals or any equivalent heteroatom species.

5.7 ORGANIC REDOX REACTIONS

The nominal oxidation numbers defined in Chapter 2 can help to find reduction/oxidation (**redox**) reactions. If you check changes to the nominal oxidation number, you will see that redox reactions are equivalent to addition/elimination processes. See Figure 5.11.

Oxidation is usually the elimination of the elements of H_2 from between two bonded atoms. Because a multiple bond is formed, there is an increase in the bond order. Reduction is usually a decrease in bond order which is caused by the addition of H_2 to a π bond between two atoms which are connected by a multiple bond.

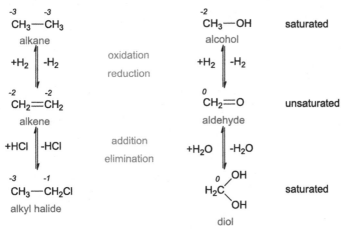

FIGURE 5.11
Relating redox and addition–elimination reactions.

We see these relationships clearly if we look at changes in the nominal oxidation numbers. For example, to change the saturated alkane CH_3–CH_3 to the unsaturated alkene CH_2=CH_2, the nominal oxidation number of each carbon changes from –3 to –2. This overall change of +2 shows oxidation. Reduction results if the reverse reaction from alkene to alkane occurs.

The same is true for the heteroatom example of alcohol conversion to aldehyde. The oxidation number of the carbon changes from −2 in the alcohol to 0 in the aldehyde, an oxidative change of +2.

Now compare the non-redox addition/elimination sequences which are shown in Figure 5.11. The elimination of H–Cl gives a sum of nominal oxidation numbers (−4) of the carbon atoms in the alkyl halide (−3 and −1). This is the same as for the product alkene (−2 and −2). Similar calculation for the addition/elimination of H_2O shows no change to the nominal oxidation number of 0 at the carbon center.

The above gives a useful general guideline to recognizing organic redox processes. This approach should not be taken to extremes for all systems.

QUESTIONS AND PROGRAMS

PROGRAM 13 Bond Breaking and Making

A   A full description of an organic reaction must include the bond breaking, bond making, and any reactive species in these processes. We can get these details from an understanding of the structural features of the molecules in the reaction.

$$
\begin{array}{c}
H \\
| \\
H\!-\!C\!-\!Br \\
| \\
H
\end{array}
$$

✎ Study the parts of the above molecule. Draw the most likely products of bond breaking.

B   The most highly polarized bond in the molecule in **A** is the C–Br. It is the most likely covalent bond to break. The bond energies are in fact C–Br 285 kJ/mol and C–H 414 kJ/mol. Therefore, you should have drawn:

$$
\begin{array}{c}
H \\
| \\
H\!-\!C\!-\!Br \\
| \\
H
\end{array}
\quad\text{------}\longrightarrow\quad
\begin{array}{c}
H \\
| \\
H\!-\!C\,\oplus \\
| \\
H
\end{array}
\quad\text{and}\quad
Br^{\ominus}
$$

This shows the most likely process, i.e., heterolysis, to give the product ions. The alternative below is less likely because of the direction of the original C–Br polarization.

$$
\begin{array}{c}
H \\
| \\
H\!-\!C\!-\!Br \\
| \\
H
\end{array}
\quad\text{------}\longrightarrow\quad
\begin{array}{c}
H \\
| \\
H\!-\!C\,\ominus \\
| \\
H
\end{array}
\quad\text{and}\quad
Br^{\oplus}
$$

✎ Draw a third possible outcome for cleavage of the C–Br bond.

C   This is the symmetrical homolysis to give radicals.

$$
\begin{array}{c}
H \\
| \\
H\!-\!C\!-\!Br \\
| \\
H
\end{array}
\quad\text{------}\longrightarrow\quad
\begin{array}{c}
H \\
| \\
H\!-\!C\cdot \\
| \\
H
\end{array}
\quad\text{and}\quad
Br\,\cdot
$$

The reverse of the above processes is heterogenic and homogenic bond making. Together, these pairs of processes are the polar and radical reaction pathways.

✎ If we ignore bond polarity, what other bond breaking can we draw?

D This is the possible heterolysis or homolysis of a C–H bond.

$$H—\overset{\overset{H}{|}}{\underset{\underset{H}{|}}{C}}—Br \quad \cdots\cdots\blacktriangleright \quad \overset{\ominus}{C}—Br \quad and \quad H^{\oplus}$$

Note that only the option in the direction of the small polarization of the C–H bond toward carbon is shown.

$$H—\overset{\overset{H}{|}}{\underset{\underset{H}{|}}{C}}—Br \quad \cdots\cdots\blacktriangleright \quad \cdot C—Br \quad and \quad H\cdot$$

The bond, C–Br or C–H, which is actually broken in the reaction will depend on the relative bond energies. in this case, the C–Br bond has the lower bond energy.

You can deal with most polar and radical reactions in this way. When similar ionic and radical reactive intermediates occur, the ease of reaction will depend on their relative stability.

Q 5.1. Name the functional groups in the following molecules and show the direction of the polarization within those functional groups.

Q 5.2. Name the species or bolded atomic center as either electrophilic or nucleophilic.

$$H^{\oplus} \quad {}^{\ominus}OH \quad Br^{\oplus} \quad CH_3—OH \quad NH_3 \quad (CH_3)_3C^{\oplus}$$

$$CH_3—C\equiv C—H \quad \overset{H}{\underset{H}{C}}=O \quad CH_3—C\equiv N$$

Q 5.3. Explain the term "polar reaction."

PROGRAM 14 Polar Reaction Types

A Although polar reactions occur in different reaction types, the basic features of a polar reaction are the same.

$${}^{\ominus}OH \quad + \quad CH_3—Br \quad \longrightarrow \quad CH_3—OH \quad + \quad Br^{\ominus}$$

✎ Try to pick out these features in the above substitution reaction.

B You should have drawn something like.

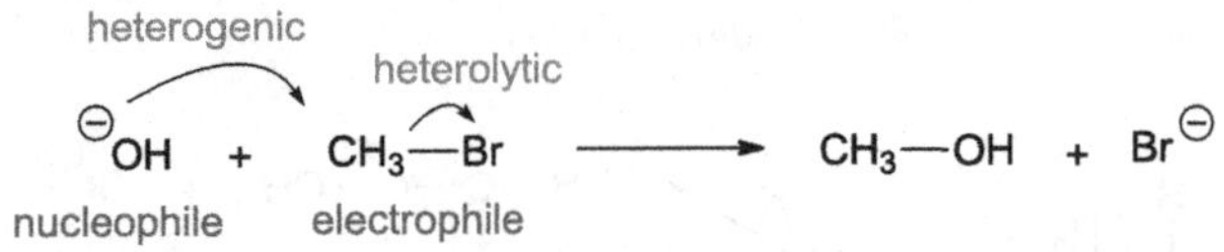

heterogenic

heterolytic

$^{\ominus}$OH + CH$_3$—Br ⟶ CH$_3$—OH + Br$^{\ominus}$

nucleophile   electrophile

This shows the electron pair of the hydroxide anion making a heterogenic bond with the electrophilic carbon which is caused by inductive polarization. At the same time, the heterolytic loss of bromide occurs.

The reaction is a substitution because only a simple exchange of one single bond for another occurs. There is no change in bond order or hybridization state at the atomic centers.

✎ What happens when the non-ionic nucleophile below reacts with an electrophile such as a proton?

CH$_2$=CH$_2$ + H$^{\oplus}$ ----?----▶

C For this polar reaction to occur, a pair of electrons from the C=C nucleophile is shared with the electrophile.

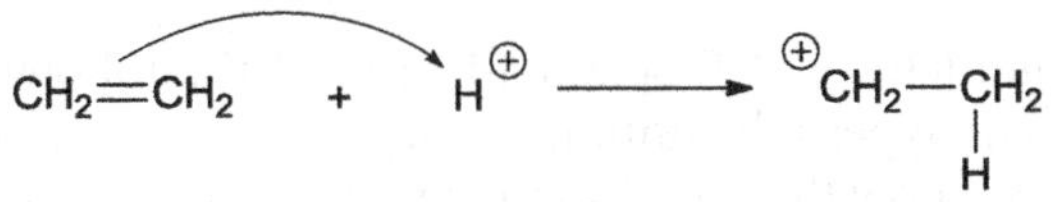

CH$_2$=CH$_2$ + H$^{\oplus}$ ⟶ $^{\oplus}$CH$_2$—CH$_2$
                                              |
                                              H

The relatively weakly held π-bond electron pair is used to make the new C–H bond. The double bond is lost and the remaining carbon, which is one electron short, is a carbocation. This high energy reactive intermediate needs a nucleophile to finish the overall reaction.

✎ Can you suggest a suitable nucleophile?

D One example could be a bromide ion. Then the overall reaction becomes:

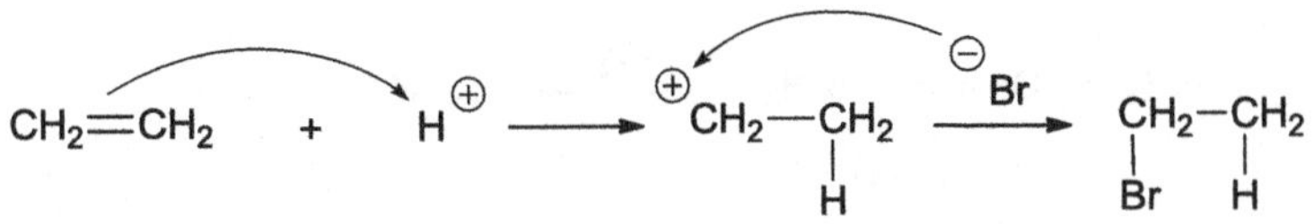

CH$_2$=CH$_2$ + H$^{\oplus}$ ⟶ $^{\oplus}$CH$_2$—CH$_2$ Br$^{\ominus}$ ⟶ CH$_2$—CH$_2$
                                              | | |
                                              H   Br   H

This reaction has two steps and results in a decrease in the original bond from double to single. Tetravalent carbon requires that two new single bonds, one on each carbon, are formed.

This is, of course, the description of an addition reaction. For this change in bond order to occur, the hybridization state of the carbon centers must change. In this case it is from sp^2 to sp^3.

✎ What is the polar reaction if the product alkyl halide is treated with hydroxide?

E   One choice is a nucleophilic substitution similar to the one described earlier - pathway A. However, alternate pathway B could be followed.

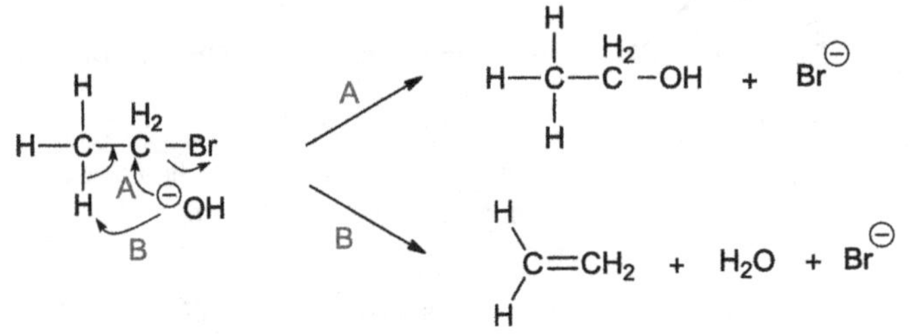

Pathway B shows the attack of the hydroxide at one of the hydrogen atoms on the carbon next to the alkyl halide. The loss of this hydrogen as a proton leaves the electron pair behind on the carbon. Then this electron pair gives a double bond between the two carbon atoms by displacing the bromide ion.

The overall result is the reverse of the addition described in **D** and is an elimination reaction. This causes an increase in the bond order from single to double, and a change in hybridization at the carbons from sp^3 to sp^2.

Q 5.4.   Use the guidelines of Program 14 to label the following reactions by their general type: substitution, addition, elimination. Do not worry if you do not know the actual specific reaction.

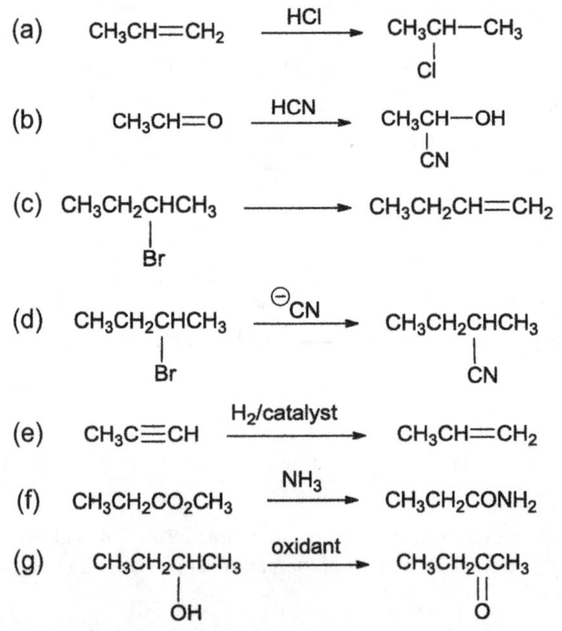

(a) $CH_3CH{=}CH_2$ $\xrightarrow{\text{HCl}}$ $CH_3CH{-}CH_3$
                                             $|$
                                             Cl

(b) $CH_3CH{=}O$ $\xrightarrow{\text{HCN}}$ $CH_3CH{-}OH$
                                           $|$
                                           CN

(c) $CH_3CH_2CHCH_3$ $\longrightarrow$ $CH_3CH_2CH{=}CH_2$
            $|$
            Br

(d) $CH_3CH_2CHCH_3$ $\xrightarrow{\ominus CN}$ $CH_3CH_2CHCH_3$
            $|$ $|$
            Br CN

(e) $CH_3C{\equiv}CH$ $\xrightarrow{\text{H}_2\text{/catalyst}}$ $CH_3CH{=}CH_2$

(f) $CH_3CH_2CO_2CH_3$ $\xrightarrow{\text{NH}_3}$ $CH_3CH_2CONH_2$

(g) $CH_3CH_2CHCH_3$ $\xrightarrow{\text{oxidant}}$ $CH_3CH_2CCH_3$
            $|$ $\|$
            OH O

Q 5.5. From the following sets of reactive intermediates, choose the one which is likely to be the most stable. Give reasons for your choices?

(a) $CH_3CH_2-\overset{\oplus}{C}H_2$ $CH_3-\overset{H}{\underset{\oplus}{C}}-CH_3$

(b) $CH_3-\overset{H}{\underset{\oplus}{C}}-CH_3$ $CH_3-\overset{H}{\underset{\oplus}{C}}-\overset{}{\underset{H}{C}}=CH_2$

(c) $CH_3-O^{\ominus}$ $\overset{CH_3}{\underset{CH_3}{\diagup}}CH-O^{\ominus}$

(d) $(CH_3)_3C\cdot$ $CH_3\overset{\cdot}{C}H_2$

(e) $CH_3C\overset{\diagup O}{\underset{\diagdown}{\ominus CH_2}}$ $CH_3C\overset{\diagup CH_2}{\underset{\diagdown}{\ominus CH_2}}$

(f) $CH_3-\overset{H}{\underset{\oplus}{C}}-CH_2CH_3$ $\oplus CH_2-CH_2CH=CH_2$

PROGRAM 15 Reaction Mechanism

A   The idea of a reaction mechanism is often difficult for those starting their study of organic chemistry. It is true that some mechanisms can seem complicated. However, the basic mechanisms of common reactions are a simple series of steps that use reactive centers, reactive intermediates, and curly arrows.

Programs 13 and 14 show bond breaking and making in simple substitution, addition, and elimination reactions. Some of these descriptions show how the reactive intermediates, radicals, carbocations, and carbanions are formed.

If you understand transition states and reactive intermediates, it is easier to write reaction mechanisms.

To practice this, study again the simple electrophilic addition reaction shown earlier in Program 14C.

$$CH_2{=}CH_2 \quad + \quad \overset{\oplus}{H} \longrightarrow \overset{\oplus}{C}H_2-\underset{H}{\underset{|}{C}}H_2$$

This first step shows the formation of a carbocation reactive intermediate. This process must have gone through a transition state.

✎ Try to show what this might look like.

B   The energy maximum which shows the transition state must occur at a point between the breaking of the π bond and the formation of the new C–H bond. This can be drawn as:

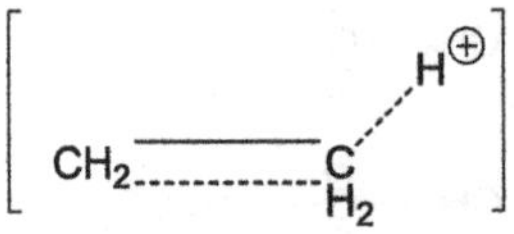

or

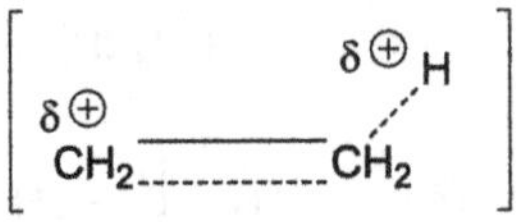

The reaction moves from this transition state to the lower energy intermediate carbocation. To finish the two-step addition, the nucleophile reacts with the sp^2-hybridized carbocation. This process goes through another transition state, with its own energy of activation, to give the product.

🖎 Show what this second transition state might look like.

C   You should have drawn something like:

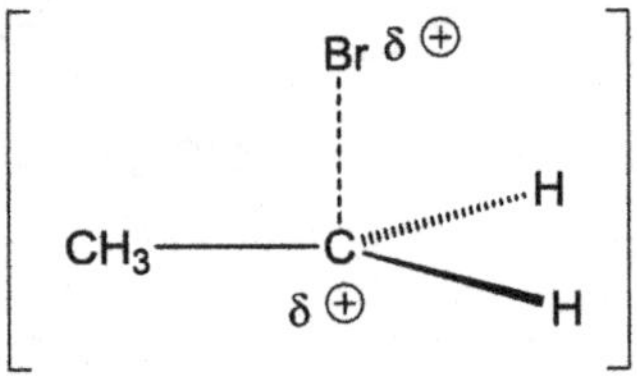

A related pair of transition states can be drawn for the reaction of a polar multiple bond such as a carbonyl group.

🖎 Try to draw these for the following reaction.

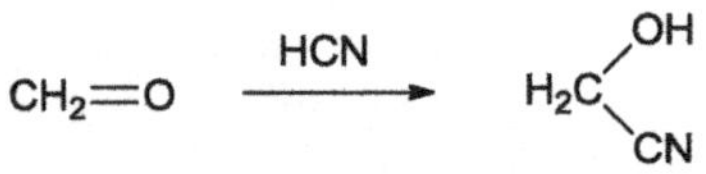

$$CH_2{=}O \quad \xrightarrow{HCN} \quad H_2C\overset{\displaystyle OH}{\underset{\displaystyle CN}{<}}$$

D   The real difference in the mechanism of this addition reaction is the order of bond formations. Because of the dipolar nature of the C=O, the CN⁻ nucleophile attacks at the electrophilic carbon to give an oxyanion intermediate. The reaction is completed by protonation of the anion to give the final addition cyanohydrin product. The transition states and the intermediate could look like:

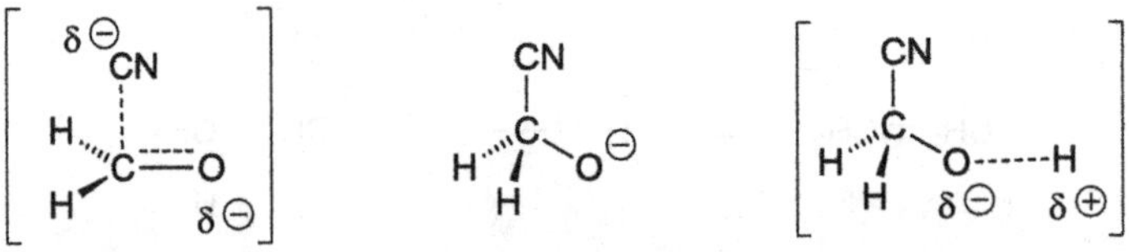

Transition
State 1

Intermediate

Transition
State 2

In the same way, you can suggest transition states for other types of reaction.

🖎 Try to draw a transition state for the nucleophilic substitution depicted in Program 14A.

$$^{\ominus}OH + CH_3{-}Br \longrightarrow CH_3OH + Br^{\ominus}$$

E   There is no evidence of a reactive intermediate, and the reaction has only one step. A suitable transition state must show that the new C–OH bond is made at the same time as the C–Br bond breaks. This can be drawn as a high energy trigonal bipyramidal 5-coordinate structure.

$$\left[\begin{array}{c} \overset{\delta\ominus}{HO} \cdots\cdots \overset{\overset{H}{\underset{|}{\overset{\cdots}{C}}}^{H}}{\underset{H}{C}} \cdots\cdots \overset{\delta\ominus}{Br} \end{array}\right]$$

Nearly all of the reactions that you see in this book are simply different versions of the same approach. Because many of them have reactive intermediates, it is clear that it is important to understand the factors which stabilize or destabilize them.

Q 5.6. Use the concepts of Program 14 to study the two reaction pathways shown in Program 14E. Draw transition states for the substitution and elimination reactions as shown.

Q 5.7. The reactions discussed in **Q 5.6** could also occur without the hydroxide ion in the first step. What differences would you now think about for the reaction pathways, transition states, and any intermediates?

Q 5.8. Find any redox reactions present in **Q 5.4**. Give nominal oxidation numbers to the carbon atoms to check the reactions that you chose.

CHAPTER 6
Acids and Bases

6.1 WHY ARE ACIDS AND BASES IMPORTANT?

In organic chemistry, the reactivity of organic acids and bases determines the reactions of many functional classes of compounds. We understand carboxylic acids as acidic from their class name. Amines are seen as basic because they are related to ammonia. This lets us understand the acid–base reactions of these two functional classes.

However, it is less easy to connect the broader ideas of acid–base reactions among other functional classes of compounds. There is a direct relationship between acidity or basicity and the concepts of electronegativity, bond polarity, inductive effect (Chapter 1), oxidation number (Chapter 2), and resonance (Chapter 4).

6.2 GENERAL DEFINITIONS

As shown in Figure 6.1, in the **Brønsted–Lowry** definition, an acid is any substance which is a donor of a proton (hydrogen ion, H^+). A base is any substance which is a proton acceptor. The reaction between an acid and a base is simply a proton transfer.

The Brønsted–Lowry definition can be used for most of the examples in this book.

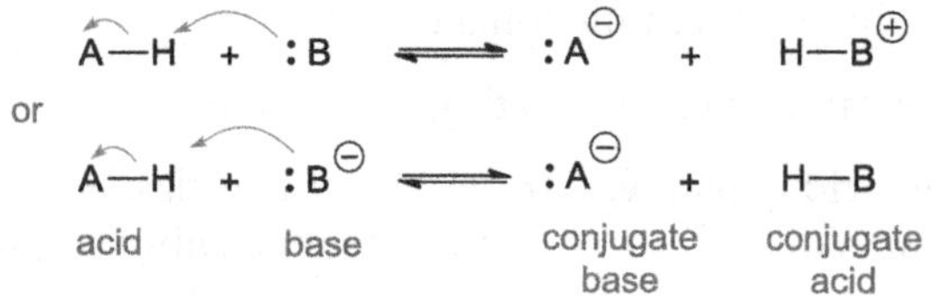

FIGURE 6.1
Brønsted–Lowry acid–base definition.

The **Lewis** definition is wider and is not limited to the transfer of a proton. As shown in Figure 6.2, an acid is a substance which accepts an electron pair, and a base donates an electron pair. A proton is a Lewis acid because it accepts an electron pair into its empty $1s$ orbital when it reacts with a Lewis base. The reaction between an acid and a base gives a Lewis acid/Lewis base complex.

Organic Chemistry Concepts: An EFL Approach. http://dx.doi.org/10.1016/B978-0-12-801699-2.00006-7

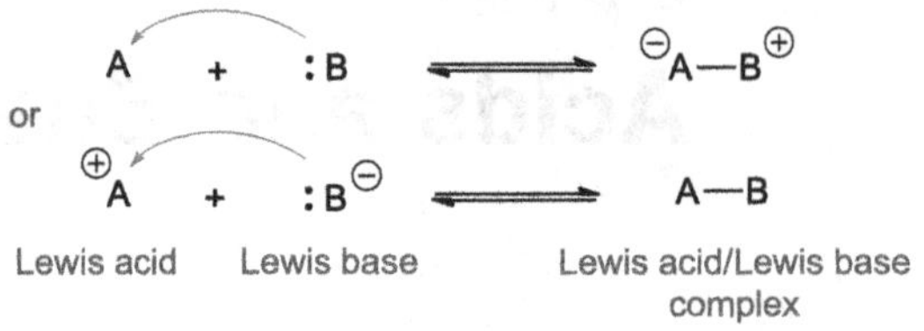

FIGURE 6.2
Lewis acid–base definition.

Figure 6.3 shows how in Brønsted–Lowry terms, the acidity constant K_a describes the strength of any acid in water solution. This can also be written as a pK_a value, in which $pK_a = -\log K_a$. Therefore, a stronger acid has a larger K_a and a lower pK_a. A weaker acid has a smaller K_a and a higher pK_a.

$$HA + H_2O \rightleftharpoons H_3O^{\oplus} + A^{\ominus}$$

$$K_{eq} = \frac{[H_3O^{\oplus}][A^{\ominus}]}{[HA][H_2O]} \qquad K_a = \frac{[H_3O^{\oplus}][A^{\ominus}]}{[HA]}$$

FIGURE 6.3
Expression of acidity.

There is an inverse relationship between the acid strength and the base strength of its conjugate base. If an acid is strong, its conjugate base does not bind the proton strongly and so it is a weak base. Especially in organic compounds, it is the structure and stability of the conjugate species which has the greatest effect on acid or base strength.

6.3 ACIDITY OF CARBOXYLIC ACIDS

Carboxylic acids are a good example for a general discussion of acidity. They show clearly the role of the electronic factors which you need to study when looking at the acidity of other functional classes.

The two aspects of carboxylic acids to think about are:

- their higher acidity relative to other functional classes;
- the differences of acidity over many different carboxylic acids.

$$\underset{\text{carboxylic acid}}{RCO_2H} \rightleftharpoons \underset{\text{carboxylate conjugate base}}{RCO_2^{\ominus}} + H^{\oplus}$$

$$K_a = \frac{[H^{\oplus}][RCO_2^{\ominus}]}{[RCO_2H]} \quad \text{and} \quad pK_a = -\log K_a$$

FIGURE 6.4
Acidity of carboxylic acids.

The position of the equilibrium in Figure 6.4 gives the degree of acidity. This equilibrium is controlled by how easily the proton is lost by the carboxylic acid. The factors which control this proton loss include:

- the initial strength of the O–H bond. This depends on the polarity of the bond.
- the stability of the carboxylate anion conjugate base. This depends on how well it carries the negative charge which results.

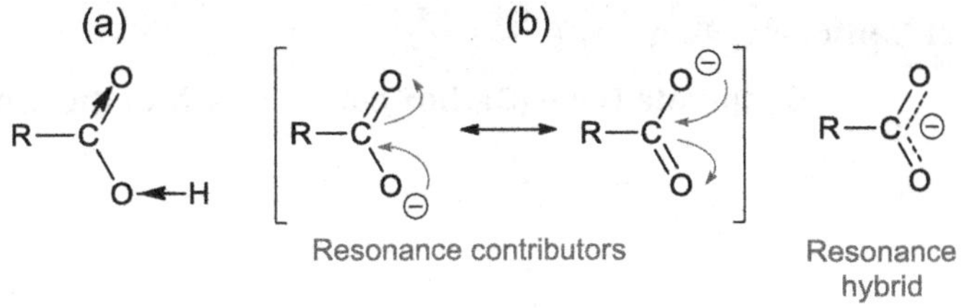

FIGURE 6.5
Factors affecting acidity.

As seen in Figure 6.5(a), the carboxylic acid structure shows these factors especially well. The O–H bond is highly polarized because of the high electronegativity of oxygen. This makes the bond weaker. The carbonyl group (C=O) increases the overall electron withdrawal.

The carboxylate anion is stabilized because:

- the negative charge is carried by the highly electronegative oxygen;
- the negative charge is delocalized over the two equal resonance forms shown in Figure 6.5(b). This means each oxygen atom has only half the charge. The resonance hybrid has a lower energy than either of the resonance forms. We can prove this if we measure the C–O bond lengths in carboxylate salts. Both C–O bonds are equal. They have bond lengths which are between typical single and double carbon–oxygen bond lengths.

All carboxylic acids have the same two features mentioned earlier. Why are some more acidic than others?

Table 6.1 shows the role which the inductive effect of the R-group has on the degree of acidity. This depends on the relative electron donation or withdrawal by substituents on the α-carbon. In other words, the carbon is bonded directly to the carboxyl group. As shown in Chapter 2, the carboxyl carbon has a nominal oxidation number of +3. This is usually more electronegative than the α-carbon with oxidation number –3 to 0 for carbon in a 1° to 4° alkyl group.

If the α-carbon does not have electronegative groups, for example halogens, it has a positive inductive effect. This +I electron donating effect decreases the polarity of the O–H bond and the acidity. Any donation of electron density also makes the carboxylate anion less stable. In other words, it causes a stronger conjugate base.

If there are electronegative groups on the α-carbon, there is an overall –I effect. The electron withdrawing effect increases the polarity of the O–H bond and the acidity. Any withdrawal of the negative charge also makes the carboxylate anion more stable. In other words, the conjugate base is weaker.

The inductive effect of the R-group of a carboxylic acid also changes the polarization of the O–H bond in the original carboxylic acid.

Table 6.1 Relative Acidities of Carboxylic Acids

pK_a	Acid	Conjugate Base (Carboxylate)	ON of the α-carbon
3.77			-
4.76			–3
2.86			–1
1.29			+1
0.65			+3

Be careful when evaluating the inductive effect. From Chapter 1 you should remember that the inductive effect occurs through the tightly held σ bond electrons. Any inductive effect drops off beyond one bond distance, unless multiple bonding is present to extend the effect through conjugation. Therefore, if the electron withdrawing groups are not on the α carbon, they have little or no effect on acidity.

So far, we have given simple guidelines to explain the relative acidity of carboxylic acids as a functional class. We can use the same general ideas of X–H bond strength and the stability of the conjugate base anion when we study other functional classes.

6.4 GENERAL FUNCTIONAL GROUP ACIDITY

Carboxylic acids are generally the most acidic of the organic compounds. However, because nearly all organic compounds have hydrogen, most other classes of compounds can show some acidity. It is possible to draw a scale of acidity which covers all classes of organic compounds. However, only some of these classes have acidities which are in the useful range.

There is an inverse relationship between substrate acidity and the basicity of the deprotonated conjugate. Therefore, it is also possible to make a scale of organic bases. However, it is more useful to have all organic compounds on a common pK_a scale. This scale shows the ability of compounds to donate or accept protons, either in their protonated acidic or deprotonated conjugate base forms. The weaker the acid, the stronger the conjugate base.

We must carefully study functional groups when we try to determine acidity. For example in Figure 6.6, the overall functional group in a carboxylic acid is shown as a combination of the carbonyl and the alcohol functional groups. Both of these functional groups affect the acidity of a carboxylic acid. In other compound types, each functional group must be evaluated for its effect on acidity.

FIGURE 6.6
The origins of the carboxyl group.

The examples in Table 6.2 show the wide range of pK_a values. At one end of the scale, carboxylic acids are the most acidic. They have the lowest pK_a values and have the weakest conjugate bases. This shows the efficiency of the combination of inductive effect and charge delocalization.

At the other end of the scale, the hydrocarbons alkanes, alkenes, and alkynes have high pK_a values. This shows that without either inductive effect or charge delocalization, there is extremely low acidity. Within the range of hydrocarbons, the higher the bond order of the carbon at the acidic C–H, the greater the acidity. This causes alkynes, with $pK_a \approx 25$, to be in the useful range.

Table 6.2   Selected Functional Group Acidities

Compound Class	Substrate	Conjugate Base	pK_a
Alkane	CH_4	$^{\ominus}CH_3$	48
Alkene	$H_2C{=}CH_2$	$H_2C{=}CH^{\ominus}$	44
Alkyne	$HC{\equiv}CH$	$HC{\equiv}C^{\ominus}$	25
Carbonyl	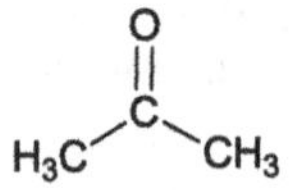 $\underset{H_3C}{}\overset{O}{\underset{}{C}}CH_3$	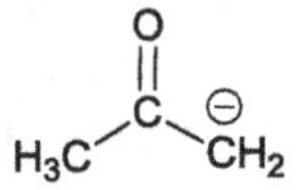 $\underset{H_3C}{}\overset{O}{\underset{}{C}}CH_2^{\ominus}$	20
Alcohol	CH_3CH_2OH	$CH_3CH_2O^{\ominus}$	16
Phenol	PhOH	$PhO^{\ominus}$	10
Carboxylic acid	CH_3CO_2H	$CH_3CO_2^{\ominus}$	4.7

6.4.1 Acidity and Hybridization

Chapter 5 showed that when a hydrocarbon loses a proton, a carbanion is formed as the conjugate base. There is no inductive effect to weaken the C–H bond, and there is no possible inductive or resonance stabilization of the conjugate carbanion. Therefore, these carbanions are very strong bases.

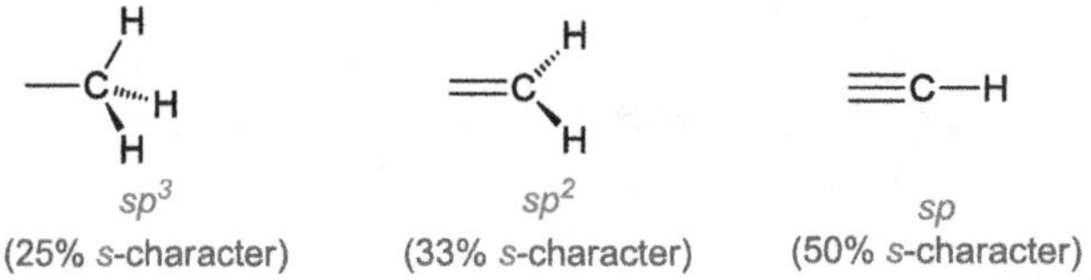

sp^3
(25% s-character)

sp^2
(33% s-character)

sp
(50% s-character)

FIGURE 6.7
Acidity and hybridization in hydrocarbons.

Figure 6.7 shows the change in s-character from sp^3 to sp hybridization states. There is an increase in s-character across the range 25–50%. We think of the s-orbital as closer to the atomic nucleus. Therefore, the greater the s-character of the orbital, the more closely bound the associated electrons. In other words, the electronegativity of carbon increases from sp^3 to sp.

Another way to look at this change in acidity is by using nominal oxidation numbers (Chapters 2 and 5). The carbon atoms in Figure 6.7 have oxidation numbers of –3 (sp^3), –2 (sp^2), and –1 (sp). This shows that an sp^3 carbon is the most reduced, least electronegative, species. Therefore, an sp carbon is the most oxidized, most electronegative, species.

6.4.2 Tautomerism and Enolization

The next most acidic compounds in Table 6.2 are those which have a carbonyl group. Earlier we showed that the carbonyl group is part of what causes the acidity in carboxylic acids. The acidity of carbonyl compounds shows the effect of the carbonyl group on acidity independent of the polar O–H bond.

The common carbonyl functional classes, aldehydes and ketones, are carbon equivalents of carboxylic acids. In carbonyl compounds, the acidic proton is on a carbon α to the carbonyl group, rather than on the oxygen of an OH group. The acidity of any protons on the α-carbon is a direct result of polarization of the C–H bond which is caused by the negative inductive effect of the carbonyl C=O group.

An example of this acidity is drawn in Figure 6.8. This shows the equilibrium between structural isomers in which the difference is simply the position of acidic hydrogen. This type of equilibrium is called **tautomerism**, and the isomers are tautomers. The exact position of the equilibrium measures the extent of **enol** tautomer formation. This shows the acidity of the carbonyl compound, and depends on the electronic and steric factors in the carbonyl compound. Tautomerism is a true equilibrium of real species, and not an example of resonance.

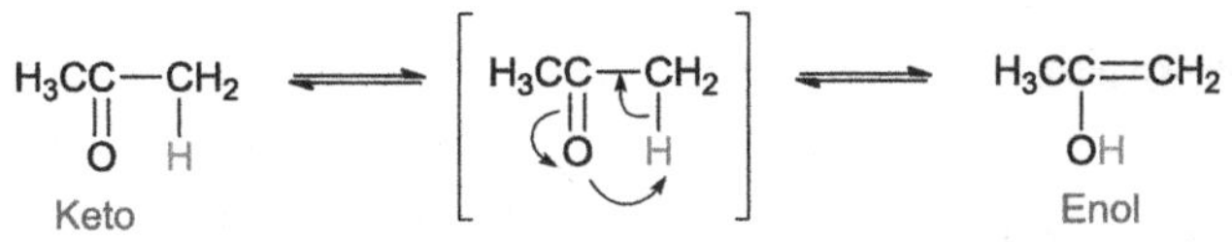

FIGURE 6.8
Tautomerism in a ketone.

The acidity of the protons on the α-carbon of carbonyl compounds makes it possible to prepare carbanions next to the carbonyl group. This forms the basis of much of the chemistry of carbonyl compounds. Figure 6.9 shows the reaction of this type of carbon acid with a base. **Deprotonation** with the base removes a proton and forms the **enolate** conjugate base. This enolate can be drawn as a resonance-stabilized anion in which the negative charge is shared between the α-carbon and the carbonyl oxygen.

The enolate is an example of resonance in which the resonance forms make unequal contributions to the resonance hybrid. The resonance form with the charge on the electronegative oxygen has lower energy and is more stable. Therefore, this resonance form makes a larger contribution to the overall hybrid. **Protonation** of the enolate in Figure 6.9 places a hydrogen atom either on the carbon or oxygen anion. This gives either the original carbonyl compound or the enol. Then the carbonyl compound and the enol return to the equilibrium as in Figure 6.8.

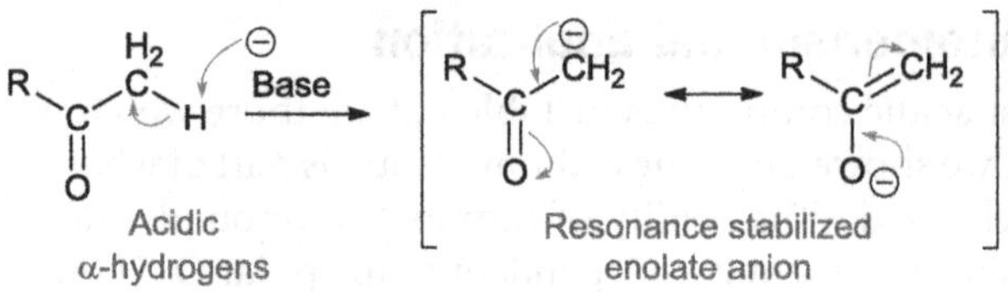

FIGURE 6.9
Enolate formation.

6.4.3 Alcohols and Phenols

Table 6.2 shows that alcohols and phenols have acidities between carbonyl compounds and carboxylic acids. Both alcohols and phenols have a highly polarized O–H bond. When the acidic proton is lost, oxyanion conjugate bases are formed. These conjugate bases are called alcoholates or alkoxides and have the negative charge on the electronegative oxygen as shown in Figure 6.10.

FIGURE 6.10
Acidity of the hydroxyl group.

The polar nature of the O–H bond explains the relative acidity of this functional group. However, phenols are relatively much more acidic than general alcohols. This is because the phenoxide conjugate base has the oxyanion directly bonded to an aromatic ring. The charge is delocalized into the ring by resonance, as seen in Figure 6.11. This makes the conjugate base more stable and of lower energy. With other alcohols, the negative charge is localized on the oxygen atom.

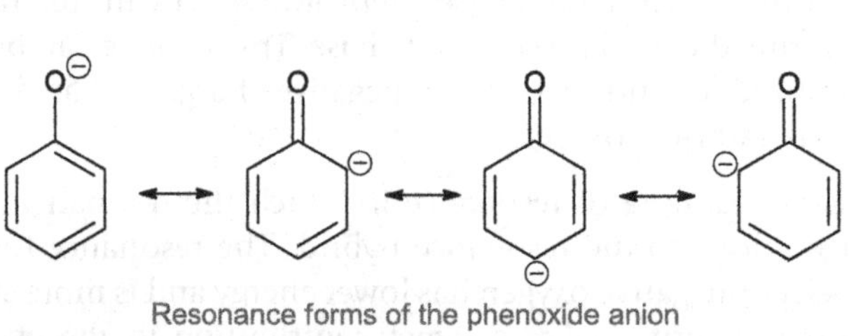

Resonance forms of the phenoxide anion

FIGURE 6.11
Resonance stabilization of phenoxide anions.

6.5 GENERAL FUNCTIONAL GROUP BASICITY

In the Brønsted–Lowry definition, we compare organic substrates based on their ability to act as bases and accept a proton. In the examples of acidity mentioned earlier, we saw different conjugate bases which are caused by deprotonation of the corresponding acids. Because acidity and basicity have an inverse relationship, the weaker the acid (higher pK_a), the stronger the conjugate base.

Therefore, alkanes give extremely strong conjugate bases because of their very low acidity ($pK_a \approx 50$). The alkoxides, from alcohols ($pK_a \approx 16$), are in fact commonly used as relatively strong bases in a wide range of organic reactions.

Organic compounds in their protonated form can also act as bases. To do this, they need to have a functional group which can provide the pair of electrons which is needed to bind an additional proton. The classification of organic bases is almost the same as in the Lewis definition. In other words, they must be able to donate a pair of electrons. In this book we only show the examples in which this electron pair is shared with a proton.

This means that all organic compounds which have heteroatoms such as O, N, S, or halogen can be basic. This is because they all have lone pairs of non-bonded electrons. Amines, alcohols, and ethers are the most common examples.

In most cases, the reaction of a lone pair as a base simply means protonation to give the charged species such as the ammonium or oxonium. As a measure of base strength, we define the constant K_b in the same way as for K_a. This is shown in Figure 6.12. Again, it is more convenient to express this as pK_b in terms of $-\log K_b$. As with pK_a, the lower the value of pK_b, the stronger the base.

$$RX: \; + \; H_2O \; \rightleftharpoons \; \overset{\oplus}{RXH} \; + \; \overset{\ominus}{OH}$$

$$K_b = \frac{[\overset{\oplus}{RXH}][\overset{\ominus}{OH}]}{[R\overset{..}{X}]} \quad \text{and} \quad pK_b = -\log K_b$$

FIGURE 6.12
Expression of basicity.

To allow direct comparison on a single scale, it is common to express pK_b values in terms of the pK_a of the conjugate acids. Because of the simple inverse relationship, we get values from the relationship $pK_a + pK_b = 14$. Selected values are listed in Table 6.3. This gives a single scale of the abilities of organic compounds to donate or accept protons.

Table 6.3   Selected Functional Group Basicities

Compound Class	Substrate	Conjugate Acid	pK_a
Carbonyl	$(CH_3)_2C{=}O$	$(CH_3)_2C{=}\overset{\oplus}{O}H$	−7
Phenol	$PhOH$	$\overset{\oplus}{Ph}OH_2$	−6.4
Ether	$(CH_3)_2O$	$(CH_3)_2\overset{\oplus}{O}H$	−3.5
Alcohol	CH_3CH_2OH	$CH_3CH_2\overset{\oplus}{O}H_2$	−2.4
Water	H_2O	$H_3O^{\oplus}$	−1.7
Ammonia	NH_3	$\overset{\oplus}{N}H_4$	9.2
Amines	$CH_3CH_2NH_2$	$CH_3CH_2\overset{\oplus}{N}H_3$	10.6

6.5.1   Basicity and Nucleophiles

In Chapter 5, we introduced the general idea of a nucleophile. A study of ionic and lone-pair nucleophiles shows they are mostly the same species as the bases which we discussed earlier. In many cases, the terms nucleophile and base seem to be the same. In other words, nucleophiles act as bases and bases act as nucleophiles.

However, because of various other factors, it is not always true that a strong base is a good nucleophile. In this book, we have mostly used the word base when the reaction takes place with a proton.

QUESTIONS AND PROGRAMS

Q 6.1. Mark each of the following species as an acid and/or base. Then draw the conjugate acid or conjugate base for each.

$H_3O^{\oplus}$ H_2O $CH_3S^{\ominus}$ NH_3 CH_3OH CH_3NH_2

Q 6.2. Write the following as proton transfer reactions. Use curly arrows to show electron flow, and label the acids, bases, and their conjugates.

(a) CH_3OH + $\overset{\oplus}{N}H_4$ ⟶ $CH_3\overset{\oplus}{O}H_2$ + NH_3

(b) CH_3OH + $\overset{\ominus}{N}H_2$ ⟶ $CH_3\overset{\ominus}{O}$ + NH_3

(c) $(H_3C)_2C{=}O$ + $H_3\overset{\oplus}{O}$ ⟶ $(H_3C)_2C{=}\overset{\oplus}{O}H$ + H_2O

PROGRAM 16 Acid–Base Reactivity

A With strong mineral acids, complete ionization in water can be shown as:

$$HCl + H_2O \longrightarrow Cl^{\ominus} + H_3O^{\oplus}$$

Stronger Stronger   Weaker Weaker
acid   base   base   acid

Organic compounds are weak acids or bases, and are only slightly ionized. This is better shown as an equilibrium. The relative strengths of the acid, base, and their conjugates control the position of the equilibrium.

$$HA + B^{\ominus} \rightleftharpoons BH + A^{\ominus}$$

Acid   Base   Conjugate Conjugate
                        acid   base

✍ Write down and describe the two equilibrium positions that could exist.

B You should have written down something similar to the following.

$$HA \quad + \quad B^{\ominus} \quad \xrightarrow{\quad\quad} \quad BH \quad + \quad A^{\ominus}$$

| Stronger acid | Stronger base | | Weaker acid | Weaker base |

Equilibrium favoured

or

$$HA \quad + \quad B^{\ominus} \quad \xleftarrow{\quad\quad} \quad BH \quad + \quad A^{\ominus}$$

| Weaker acid | Weaker base | | Stronger acid | Stronger base |

Equilibrium favoured

✎ Where does the equilibrium lie in the following reaction?

$$CH_3CO_2H \quad + \quad \overset{\ominus}{CN} \quad \rightleftharpoons \quad CH_3CO_2^{\ominus} \quad + \quad HCN$$

pK_a 4.76   pK_a 9.31

C An acid always loses a proton to the conjugate base of any weaker acid with a higher pK_a. Ethanoic acid is the stronger acid with a pK_a of 4.76 compared with 9.31 for hydrogen cyanide. Therefore, the equilibrium lies toward the right.

$$CH_3CO_2H \quad + \quad \overset{\ominus}{CN} \quad \rightleftharpoons \quad CH_3CO_2^{\ominus} \quad + \quad HCN$$

✎ What will happen in the reaction of ethyne with hydroxide?

$$HC\equiv CH \quad + \quad \overset{\ominus}{OH} \quad \xrightarrow{\quad?\quad} \quad HC\equiv C^{\ominus} \quad + \quad H_2O$$

pK_a 25   pK_a 15.74

D H_2O is the stronger acid with the lower pK_a, and the reaction will not occur as it is drawn. Instead, the reverse reaction will occur. This is because the acetylide carbanion is a stronger base than hydroxide.

There is another way to decide on acid–base reactivity. For a reaction to be spontaneous, the products must be more stable (less reactive) than the reactants. See Program 17.

Q 6.3. The amide ion (H_2N^-) is a stronger base than hydroxide. Which of the conjugate acids, NH_3 or H_2O, is stronger?

Q 6.4. Ammonia has pK_a 33 and propanone has pK_a 20. Can the following reaction occur as drawn?

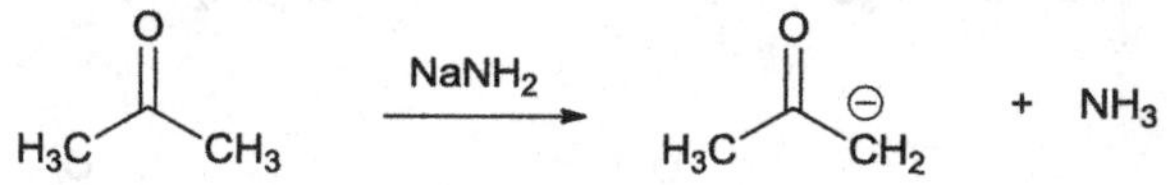

Q 6.5. Can the bicarbonate anion deprotonate methanol as shown in the following equation?

$$CH_3OH + HCO_3^{\ominus} \xrightarrow{\quad ? \quad} CH_3O + H_2CO_3$$

pK_a 15.5 \hspace{3cm} pK_a 6.4

PROGRAM 17 Acidity/Basicity and Resonance

A So far we have decided acid and base strength by using known pK_a values. If the pK_a value is not known, you can use the resonance and inductive electronic effects to explain differences in acid and base strength. This program deals with the important role of resonance.

A comparison of the acid strengths of ethanol and ethanoic acid is a good example to show the concept. To do this, we need to compare the equilibria:

$$CH_3CH_2OH \rightleftharpoons \overset{\oplus}{H} \quad CH_3CH_2O^{\ominus}$$

$$CH_3CO_2H \rightleftharpoons \overset{\oplus}{H} \quad CH_3CO_2^{\ominus}$$

A good place to start is with the possible resonance forms of the species.

Draw and compare the possible resonance structures.

B Possible resonance forms include:

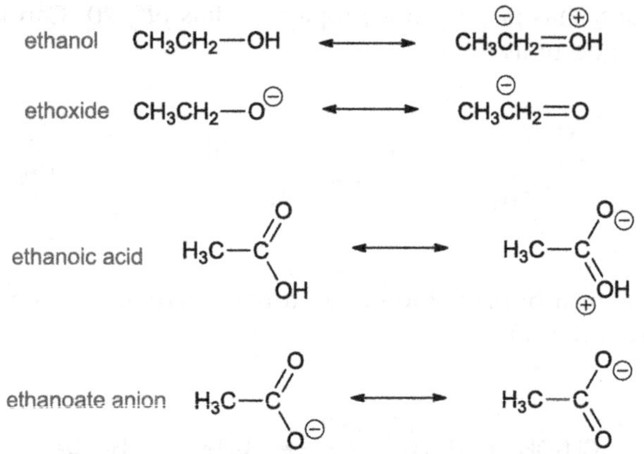

ethanol CH_3CH_2—OH ⟷ $CH_3CH_2{=}\overset{\oplus}{\underset{\ominus}{O}}H$

ethoxide CH_3CH_2—$\overset{\ominus}{O}$ ⟷ $CH_3CH_2{=}\overset{\ominus}{O}$

ethanoic acid $H_3C-C\overset{O}{\underset{OH}{}}$ ⟷ $H_3C-C\overset{O^{\ominus}}{\underset{\overset{OH}{\oplus}}{}}$

ethanoate anion $H_3C-C\overset{O}{\underset{O^{\ominus}}{}}$ ⟷ $H_3C-C\overset{O^{\ominus}}{\underset{O}{}}$

In the case of ethanol and its ethoxide conjugate base, resonance forms are not likely. This is because carbon is less likely than oxygen to have a negative charge. Therefore, we can say that neither ethanol nor the ethoxide conjugate base is resonance stabilized.

The resonance forms for ethanoic acid and the ethanoate anion are much more likely. Between these two options, the resonance forms for the ethanoate anion are better. This is because these resonance forms are equal and the resonance is symmetrical. One of the resonance forms for ethanoic acid has charge separation, and this is a relatively high energy form. The presence of resonance stabilization in ethanoic acid (pK_a 4.7) shows why it is a stronger acid than ethanol (pK_a 16).

The same method can be used with base strength. Study the example of aniline (*phenylamine*) and its saturated counterpart cyclohexylamine.

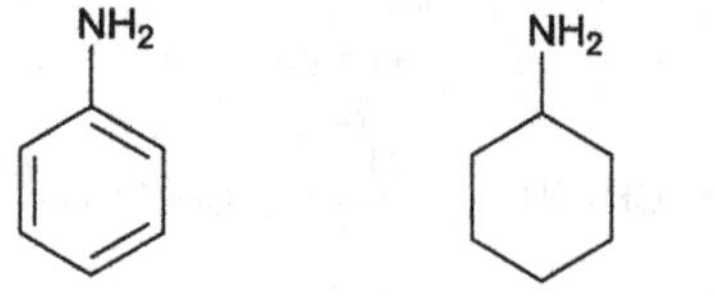

Aniline   Cyclohexylamine

✍ Use possible resonance forms to show which you expect to be the stronger base?

C To compare these molecules, you should have drawn resonance forms similar to:

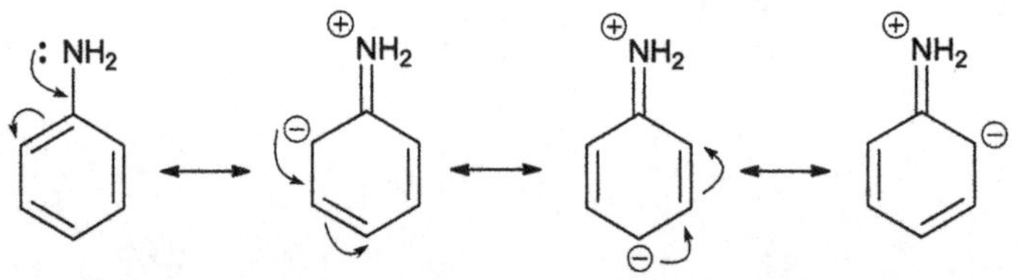

We can easily draw resonance forms for the aromatic aniline. However, there is no similar resonance for cyclohexylamine. The delocalization stabilizes the aniline and

—Cont'd

reduces the base strength of the nitrogen lone pair. If the nitrogen in aniline is protonated, it gives a relatively acidic conjugate acid of pK_a 4.6. The lone pair is used to form the bond with the proton. This means that it can no longer be delocalized into the aromatic ring.

The cyclohexylamine has a non-delocalized lone pair. This is more basic and is easily protonated to give a very weak conjugate acid of pK_a 10.7.

pK_a 10.7 pK_a 4.6

🖎 Can you explain the use of pK_a to describe the strength of bases?

D   To make them easy to compare, both acids and bases are listed on the same pK_a scale. We defined base strength as "the acid strength of the conjugate acid after protonation of the base."

A strongly acidic conjugate acid with low pK_a means that the base is very weak. A weakly acidic conjugate acid with high pK_a means a stronger base. Therefore, the higher the pK_a of a base, the stronger the base. This is the reverse of the situation for acids.

🖎 Work out if ethylamine or ethanamide is the stronger base.

E   There is delocalization of the nitrogen lone pair in the case of ethanamide. Therefore ethylamine, which has no delocalization, is the stronger base.

$$H_3C-C \quad \longleftrightarrow \quad H_3C-C \qquad\qquad CH_3CH_2NH_2$$

pK_a 0.6 pK_a 10.8

Now we return to acidity. We can apply the above resonance method to compound classes other than the alcohol and carboxylic acid example that we studied in **A**.

🖎 Is propanone or ethanoic acid the stronger acid?

F   Here we are comparing a ketone with a carboxylic acid. We need to look at resonance forms that can make one functional group more acidic than the other. To act as an acid, the molecules must give up a proton. We need to compare the relative stability of the conjugate bases that are formed by this action.

Continued...

—Cont'd

ethanoic acid

ethanoate anion

equivalent resonance forms

propanone enolate

non-equivalent resonance forms

We can see the similarity between carboxylic acids and carbonyl compounds which have a hydrogen on the α-carbon. This is the same positional arrangement as the OH hydrogen in carboxylic acids. The resulting enolate anion is resonance stabilized, and this explains the relative acidity of propanone with pK_a 20. However, the non-equivalent resonance forms are not as efficient as the equivalent carboxylate resonance forms.

This is because one resonance form for the enolate has the negative charge on a carbon. Both the symmetrical carboxylate resonance forms have the negative charge on oxygen. Since oxygen is more electronegative than carbon, it can better carry the negative charge. So, ethanoic acid has the greater acid strength with pK_a 4.7.

✍ For propanone, we cannot draw a resonance form similar to the one for ethanoic acid. Explain this.

G   As a carbon analog of a carboxylic acid, propanone has no lone pair for delocalization into the C=O. The only way to get a pair of electrons is to lose a proton. This causes the transfer of the proton to the oxygen as shown below.

keto

enol

These two structures are not resonance forms because the atomic arrangement has changed. Instead, they are the equilibrium between two structural isomers. This equilibrium is an example of tautomerism between the keto and enol forms of propanone.

✍ Do you think that esters and amides show similar α-hydrogen acidity to the ketone in the reaction?

H The answer is no. The resonance forms give an electronic explanation. Both esters ($pK_a \approx 24$) and amides ($pK_a \approx 28$) show the same resonance forms for their α-deprotonated carbanions.

In addition, the lone pairs on the oxygen and nitrogen allow us to draw other important resonance forms. These resonance forms do not make equal contributions to the hybrid, but they compete directly with the delocalization of the carbanion. Because of this, the resonance stabilization of the carbanion is not as efficient as the resonance stabilization of the ketone enolate.

The resonance approach in this program has not been applied to other classes of compound such as hydrocarbons and alkyl halides. This is because they have no possibility of resonance stabilization. Therefore, they are very weak acids. Program 18 looks at the impact of the inductive effect on acidity and basicity.

Q 6.6. Arrange the following in order of decreasing acidity at the central methylene CH_2.

$$CH_3COCH_2COCH_3 \quad | \quad CH_3COCH_2CO_2CH_3 \quad | \quad CH_2(CO_2CH_3)_2$$

Q 6.7. Draw resonance forms for the anion from the 2,4-pentandione in **Q 6.6**.

PROGRAM 18 Acidity/Basicity and Inductive Effects

A In Program 17, we studied the importance of resonance on the acidity of organic compounds. We looked at the ability of both O–H bonds (alcohols, phenols, carboxylic acids) and the α-carbon C–H bonds (aldehydes, ketones, ester, amides) to provide a proton.

These general H–A bonds are polarized to different amounts based on the electronegativity differences of the two atoms in the bond. This is the inductive effect. The higher electronegativity of oxygen compared with carbon means that an O–H bond is more dipolar than a C–H bond.

Continued...

—Cont'd

For example, look at the inductive effect in an alcohol compared with an alkane.

$$CH_3CH_2\text{—}H \underset{}{\overset{H^{\oplus}}{\rightleftharpoons}} CH_3\overset{\ominus}{C}H_2 \qquad\qquad CH_3O\text{—}H \underset{}{\overset{H^{\oplus}}{\rightleftharpoons}} CH_3O^{\ominus}$$

pK_a 50 pK_a 15.5

The inductive polarization weakens the O–H bond more than the C–H bond. In addition, the alkoxide conjugate base with the charge on oxygen is more stable than the carbanion with the charge on carbon.

✎ What do you expect for methane (CH_4) compared to chloroform ($CHCl_3$)?

B   Chloroform has the more polarized C–H bond because it has three chlorine substituents which withdraw electrons. This same strong –I effect helps to stabilize the carbanion conjugate base. Methane, which only has hydrogen substituents, has no such –I effect to activate or stabilize.

$$CH_3\text{—}H \underset{}{\overset{H^{\oplus}}{\rightleftharpoons}} {}^{\ominus}CH_3 \qquad\qquad CCl_3\text{—}H \underset{}{\overset{H^{\oplus}}{\rightleftharpoons}} {}^{\ominus}CCl_3$$

pK_a 48 pK_a 13.6

When we estimate the role of the inductive effect on relative acidity, we must compare systems which have similar resonance effects. If these electronic effects work against each other, the resonance effect is always stronger. More often, they act in the same direction.

✎ Compare the acidity of the alcohols CH_3OH, CH_3CH_2OH, $(CH_3)_2CHOH$, and $(CH_3)_3COH$.

C   When you compare these alcohols, the only difference between them is the size of the alkyl chain. The bigger the alkyl chain, the greater the positive inductive effect. An increase in the inductive effect decreases the O–H polarization and decreases the stability of the alcoholate conjugate base. The pK_a values show this pattern.

CH_3OH CH_3CH_2OH $(CH_3)_2CHOH$ $(CH_3)_3COH$

pK_a 15.5 pK_a 16 pK_a 18 pK_a 19

✎ Where does 2,2,2-trifluoroethanol ($CF_3 CH_2 OH$) fit into the above series?

D   Trifluoromethyl as a group strongly withdraws electron density. Therefore, it has a strong –I effect. This reverses the positive inductive effect of the CH_2 toward the OH. In this case, the O–H polarization and alcoholate stability are increased. The resulting pK_a value of 12.4 shows this increased acidity.

Because the inductive effect is felt through single σ-bond electrons, its effect quickly reduces with distance.

✎ Where do you expect the alcohols $CF_3CH_2CH_2OH$ and CH_3CF_2OH to fit into the above series?

E   In $CF_3CH_2CH_2OH$, the CF_3 group is too far from the OH to change its acidity. The resulting acidity is about the same as $CH_3CH_2CH_2OH$. However, in CH_3CF_2OH, the electronegative fluorine substituents are next to the OH. Therefore, the acidity is greatly increased.

We can use this same method to compare carboxylic acids. Program 17 shows that carboxylic acids all have the same resonance forms, but inductive effects cause their pK_a values to be different.

✎ Try to explain the order of acidity shown in the following series of acids.

$$CH_3-CO_2H \qquad H-CO_2H \qquad CF_3CH_2-CO_2H \qquad ClCH_2-CO_2H$$

$$pK_a\ 4.76 \qquad\quad pK_a\ 3.77 \qquad\qquad pK_a\ 3.1 \qquad\qquad\quad pK_a\ 2.86$$

$$pK_a\ 2.5 \qquad\qquad\qquad pK_a\ 2.4 \qquad\qquad\qquad pK_a\ 1.29$$

$$CCl_3-CO_2H \qquad\qquad CF_3-CO_2H$$

$$pK_a\ 0.69 \qquad\qquad\qquad pK_a\ 0.5$$

F   As stated earlier, all of the acids have the same possible resonance forms.

The direction and size of any inductive effect of R changes the polarization of the acidic O–H bond and the electron density of the carboxylate conjugate base.

The least acidic is ethanoic acid (R=CH_3). It has a +I effect from the CH_3- group compared to the hydrogen atom of formic acid (R=H). All the other examples have increasing –I effects at the α-carbon. The halogenated examples show the effect of the number, type, and placement of the halogen substituents.

$$CF_3CH_2- \qquad ClCH_2- \qquad Cl_2CH- \qquad Cl_3C- \qquad F_3C-$$

The remaining examples show the similar –I effect of methylammonium and acetyl groups.

Q 6.8. Arrange the following compounds in order of increasing acidity. CH_3CO_2H; CH_3CH_3; $Cl_3CH_2CH_2CO_2H$; $ClCH_2CO_2H$; CH_3CH_2OH; CH_3CHO

CHAPTER 7

Functional Classes II, Reactions

7.1 FUNCTIONAL GROUP INTERCONVERSIONS

Chapters 1–3 introduced the main structural features of the important classes of organic molecules and their functional groups. Chapters 4–6 explained the concepts which affect reactivity and help you to understand the organic reactions of simple functional group conversions.

The purpose of this chapter is to apply the concepts from Chapters 1–6 by using many different reactions as examples. The equations and schemes in this chapter show how the ideas from the earlier chapters allow to you understand and predict the reactions for the different functional groups.

This book does not show these reactions leading to very complex molecules. Instead, a small number of examples are used to show the type of reaction which is expected with each of the functional groups.

Unlike most books, we do not separate the processes of Functional Group Preparations and Functional Group Reactions. These are really one and the same, because the preparation of one functional group is simply the reaction of another. If a reaction product has a different functional group from the starting reagent, it means the reaction is a method for making this functional class. This approach gives you a clearer picture of functional group relationships.

Chapter 5 is especially important because it gives the types of reaction which any functional class can undergo. You need to understand the bonding in functional groups (inductive effect, resonance, acidity/basicity) and the properties of reactive species (mainly nucleophiles and electrophiles). Then, with this knowledge, you can work out reactions rather than memorize them.

Many of the reactions which you will see in this chapter can be done with different reagents and under other conditions. Because this does not change the principle, only one example is usually shown. Figure 7.1 shows the reaction links between common functional classes. This chart does not include all possible interconversions. Although some of the conversions are shown as stepwise through intermediate functional groups, it is often possible to do these in one step.

In Figure 7.1 certain functional groups have more connections than others. For example, alkyl halides are drawn in a central position because they can react in

Organic Chemistry Concepts: An EFL Approach. http://dx.doi.org/10.1016/B978-0-12-801699-2.00007-9

several ways which lead to different functional groups. The reactions in Figure 7.1 are arranged to allow us to easily see these relationships.

7.2 ALKANES

With any compound class, if you understand the functional group bonding, you will have an idea of its reactivity. Alkanes have no clear reactive feature. They show no great bond polarity and have only strong C–C (347 kJ/mol) and C–H (414 kJ/mol) σ-bonds. This explains their very low reactivity. In fact, they are widely used as solvents. However, under special conditions they can be reacted with oxygen and chlorine.

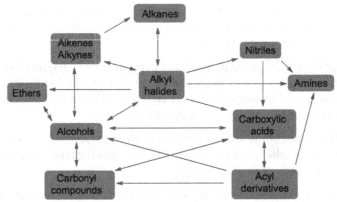

FIGURE 7.1
Selected functional group interconversions.

7.2.1 Oxidative Combustion

A **combustion** reaction occurs when saturated hydrocarbons are burned in the presence of oxygen. Examples include combustion engines or furnaces. The efficiency of combustion is controlled by the oxygen supply. If enough oxygen is supplied, the products are carbon dioxide, water, and lots of heat energy.

$$CH_4 + 2O_2 \rightarrow CO_2 + 2H_2O\,[+800\ kJ/mol]$$

$$2C_4H_{10} + 13O_2 \rightarrow 8CO_2 + 10H_2O\,[+2877\ kJ/mol]$$

If the oxygen supply is limited, the combustion is not complete. As seen in the following equations, this gives carbon monoxide and carbon deposits.

$$2CH_4 + 3O_2 \rightarrow 2CO + 4H_2O\ [\text{toxic exhaust fumes}]$$

$$CH_4 + O_2 \rightarrow C + 2H_2O\ [\text{carbon deposits}]$$

7.2.2 Radical Chlorination

In ultraviolet light, alkanes react with chlorine in a radical substitution reaction. This is one of the only radical reactions shown in this book. It is a good example to show homolytic process. The equation shows that the reaction type is a substitution in which Cl replaces H.

$$CH_4 + Cl_2 \xrightarrow{h\nu} CH_3Cl + HCl$$

The mechanism of the reaction is shown in Figure 7.2 and includes three steps:

- **initiation**—the formation of radicals needed to carry out a radical process;
- **propagation**—the repeated radical process;
- **termination**—the trapping of radicals to end a radical process.

Chapter 5 discussed the reactions in these steps and showed them as homolytic cleavage and homogenic bond making.

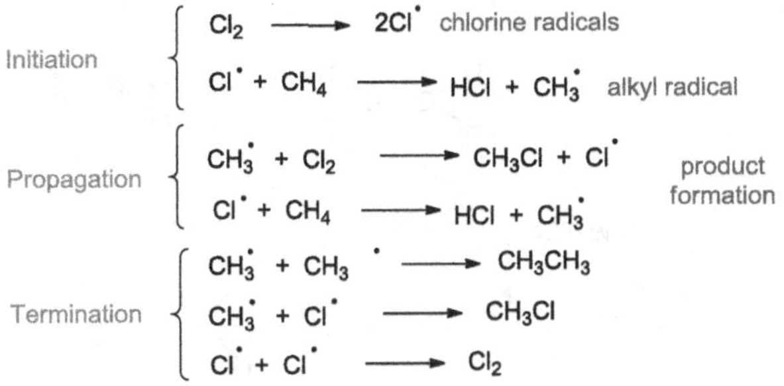

FIGURE 7.2
Stepwise description of radical substitution.

The reaction is not very practical because it gives mixtures of products. This is even more complex as the length of the alkane chain increases.

7.3 ALKENES

The unsaturated functional group (C=C) is electron rich between the bonded carbon atoms. Therefore, as shown in Chapter 5, an alkene is a potential nucleophile. The weaker held pair of π-bond electrons is reactive enough to bond with suitable electrophiles. This is the characteristic reaction of alkenes. This polar reaction in Figure 7.3 uses heterolytic bond breaking and making. Because the rate-determining step is the reaction with an electrophile, this reaction is called electrophilic addition. See Figure 7.4.

7.3.1 Electrophilic Addition

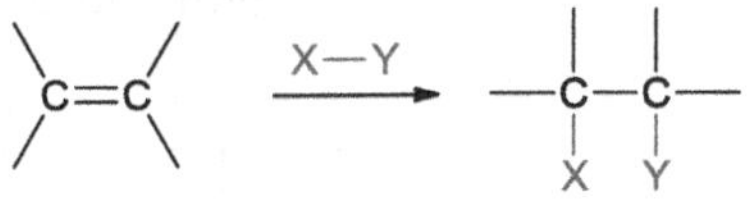

FIGURE 7.3
General electrophilic addition sequence.

As defined in Chapter 5 for an addition reaction, the bond order decreases and C=C becomes C–C. This characteristic addition to alkenes is clearly seen in the stepwise reaction mechanisms below. The reaction of symmetrical and unsymmetrical alkenes is shown separately in Figures 7.4 and 7.5.

These are our first detailed mechanisms, and they use curly arrows and reactive intermediates. Because of this, some discussion of the steps is given.

The stepwise reaction starts with heterogenic bond making between the electrophilic portion of the reagent and the pair of electrons of the π-bond. This occurs at one carbon of the original double bond. This leaves the carbon atom at the other end of the original double bond with only three ligands. This is the intermediate carbocation with only six valence shell electrons. The reaction is completed by the second heterogenic bond formation between this carbocation and the negatively charged anion portion of the reagent.

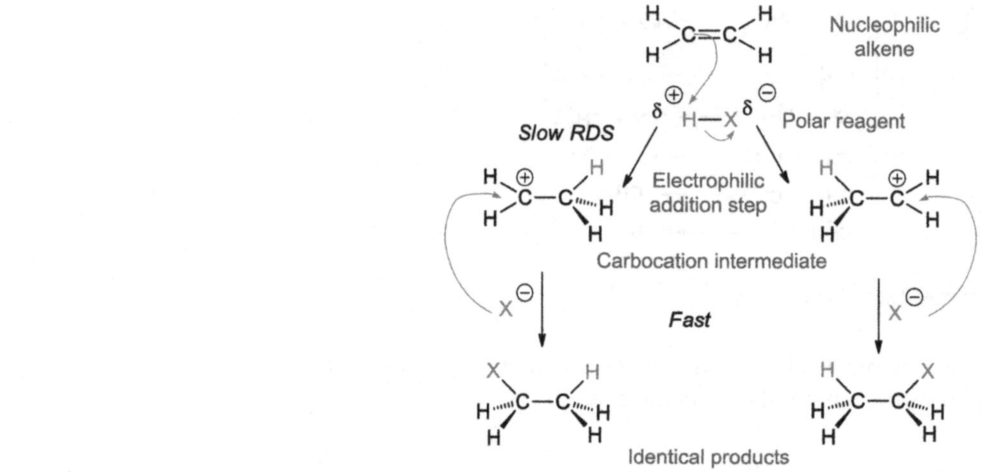

FIGURE 7.4
Mechanism of electrophilic addition to a symmetrical alkene.

In Figure 7.4, the mechanism shows it does not matter at which end of the original double bond the first bond is formed. The carbocation intermediate is the same in both cases and gives the same product. This result is because the alkene is symmetrical. Now compare this result to the unsymmetrical alkene example in Figure 7.6.

If the alkene is not symmetrical, two different products are possible when an unsymmetrical reagent is added. This is shown in Figure 7.5. Studies of the products from many of these reactions led to a finding known as the Markovnikov Rule. This can be stated simply as:

> The addition of an unsymmetrical reagent to an unsymmetrical alkene always occurs so that the electrophilic part of the reagent adds to the end of the alkene which has the most hydrogen ligands.

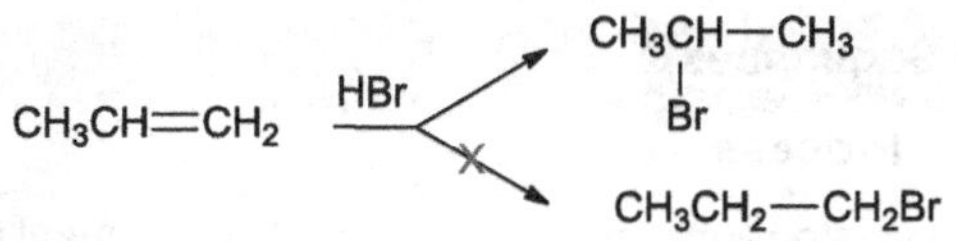

FIGURE 7.5
Markovnikov product distribution.

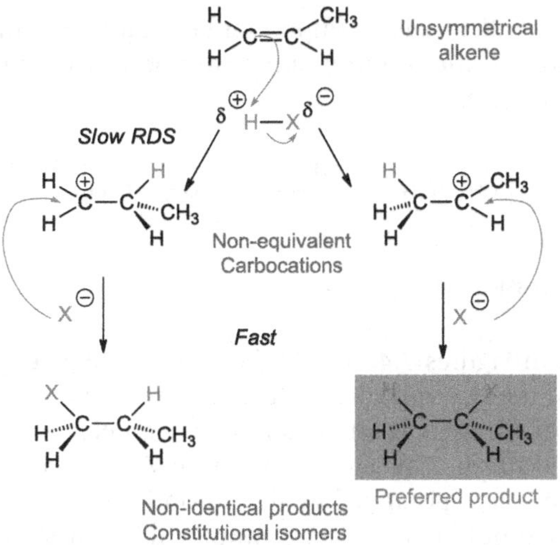

FIGURE 7.6
Mechanism of electrophilic addition to an unsymmetrical alkene.

The mechanism in Figure 7.6 explains the formation of two different products. There are two possible carbocation intermediates which can be formed by the initial reaction with the electrophile at either end of the alkene C=C. Because the alkene is not symmetrical, these carbocations are not the same and lead to different products. Because they have different structures, the carbocations do not have equal energy. Based on this, Markovnikov's rule can be restated:

> Electrophilic addition always occurs through the more stable, lower energy carbocation intermediate.

To predict the product distribution from these reactions, we need to only apply the Chapter 5 concepts which control the relative stability of carbocations.

Electrophilic addition is the main reaction of most alkene chemistry. Many different two-part reagents can be added across the alkene bond. Some common examples are listed in Table 7.1. As the table shows, these reactions are preparations of other functional groups.

Table 7.1 Common Alkene Addition Sequences

Reagent	Process	Product
HCl, HBr, HI	Hydrohalogenation	Alkyl halides
H_2O (as H_3O^+)	Hydration	Alcohols
HCN	Hydrocyanation	Alkyl cyanides (nitriles)
Cl_2, Br_2, I_2	Halogenation	Alkyl halides

The halogenation of alkenes is an important laboratory and industrial process. As shown in Figure 7.7, the addition of bromine acts as a fast visual test for the presence of unsaturation.

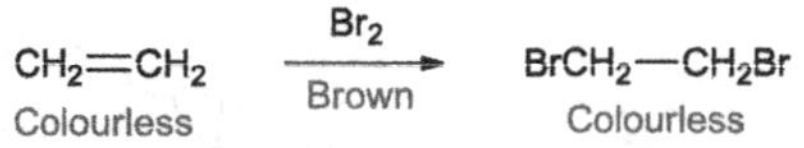
CH2=CH2
Colourless
Br2
Brown
BrCH2—CH2Br
Colourless

FIGURE 7.7
Visual chemical test for unsaturation.

The mechanisms in Figures 7.4 and 7.6 both have polar reagents in which it is easy to pick out the electrophile. In the halogenation reaction in Figure 7.7, the symmetrical diatomic reagent does not have an electrophilic end to react in the stepwise addition. The model in Figure 7.8 explains how the initial halogen molecule is first polarized by the π-bond electrons. The positive end of this temporary dipole is the electrophile in the first step of the addition reaction.

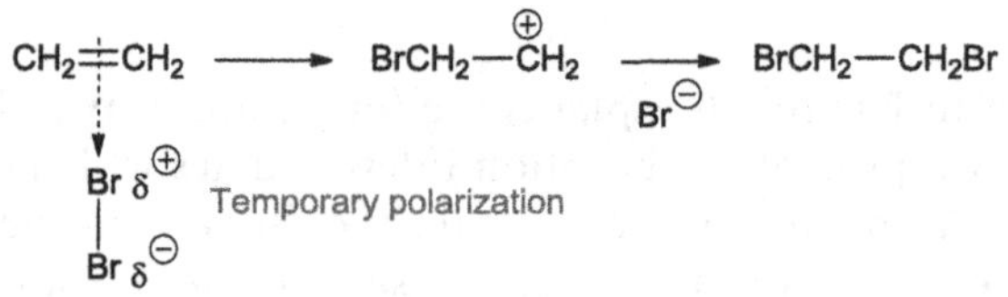
CH2=CH2
BrCH2—CH2
BrCH2—CH2Br
Br
Br δ
Temporary polarization
Br δ

FIGURE 7.8
Temporary polarization during halogenation.

7.3.2 Reduction

The addition of hydrogen was defined as reduction in Chapter 5. With alkenes, this addition is done with the help of metal catalysts. The most common catalysts are palladium on carbon (Pd-C) and platinum oxide (PtO_2).

This reduction process is called **hydrogenation** and gives a saturated product from an unsaturated one as shown in Figure 7.9. The mechanism of this reaction shows both hydrogen atoms adding to the same face of the double bond. This means there may be a specific stereochemical result when the new sp^3-hybrid carbons are formed. Although this is not needed at this point, further detail is given in Appendix 8.

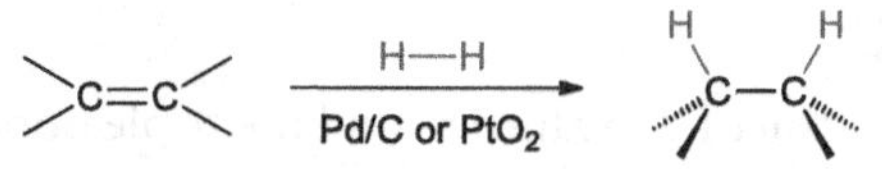

FIGURE 7.9
Hydrogenation of alkenes.

7.3.3 Oxidation

The addition of oxygen to the alkene C=C can lead to several useful products. There are three main processes: **epoxidation, hydroxylation,** and **ozonolysis.** Figure 7.10 shows these reactions and their reagents.

Epoxidation is the addition of a single oxygen atom across the alkene C=C to give a cyclic ether, which is called an epoxide. This one-step reaction is done using a peroxyacid (RCO_3H).

Hydroxylation is the addition of a hydroxyl group to each carbon of the alkene to give a 1,2-diol as the product. If the reaction uses osmium tetroxide (OsO_4) or potassium permanganate ($KMnO_4$), both hydroxyl groups are added from the same side of the double bond.

The opposite stereochemistry is obtained by **hydrolysis** of an epoxide. This decomposition of the epoxide substrate by water is acid catalyzed and gives the product in which the hydroxyl groups were added from opposite faces of the double bond.

Ozonolysis is the reaction of an alkene with ozone, O_3. This oxidation breaks both the σ- and π-bonds of the original double bond. Both carbon atoms of the C=C are changed to carbonyl C=O groups. Therefore, ozonolysis is an efficient way to prepare carbonyl compounds.

This result also shows how ozonolysis can help find the structure of complex molecules which have C=C bonds in them. The complex molecule is reacted with ozone, and the smaller carbonyl products are identified. This process helps to find the positions of the C=C bonds in the original complex structure.

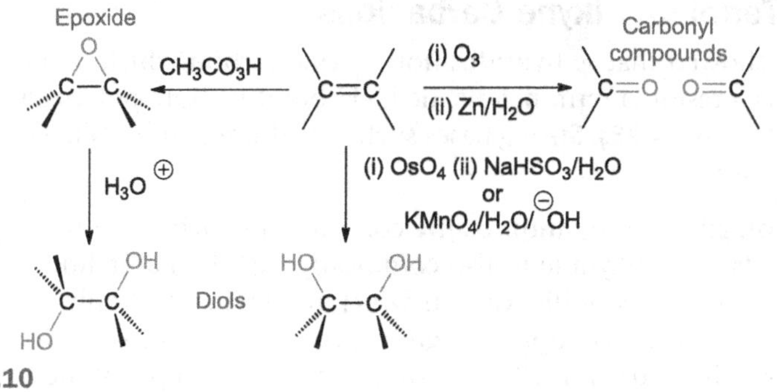

FIGURE 7.10
Oxidation reactions of alkenes.

7.4 ALKYNES

Alkynes are related in structure to alkenes, and the triple bond is a region of high electron density. For this reason, the addition chemistry of alkynes is similar to alkenes. A substitution reaction can occur with a **terminal** alkyne. Terminal means that the alkyne is at the end of the chain. One end of the alkyne is then an unsubstituted C—H group.

7.4.1 Alkyne Addition Chemistry

The electrophilic addition of hydrogen halides and halogens is as expected. Because there are two π-bonds, addition can occur two times, as shown in Figure 7.11. For unsymmetrical situations, Markovnikov's rule is followed.

Hydrogenation gives complete saturation by using two equivalents of H_2. Reduction can be stopped at the alkene stage using the Lindlar palladium catalysts, Pd-$BaCO_3$ or Pd-$CaCO_3$.

Hydration, with the help of a catalyst, initially gives alcohol products which have the hydroxy group directly attached to a C=C. This structure is the enol, and it quickly tautomerizes to the more stable ketone, as seen in Chapter 6.

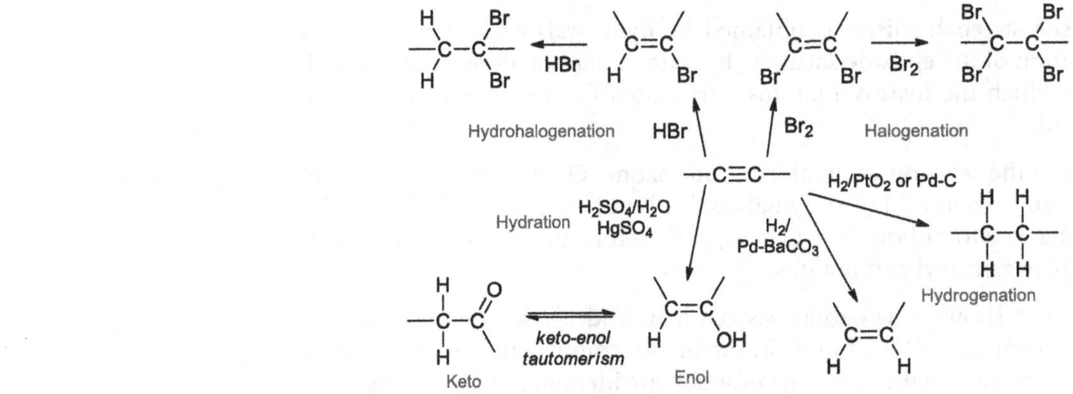

FIGURE 7.11
Alkyne addition reactions.

7.4.2 Terminal Alkyne Carbanions

Chapter 6 showed that sp-hybridization gives a carbon which is more electronegative. As a result, a terminal alkyne C–H bond is more polarized and has a higher acidity ($pK_a \approx 25$). Strong bases such as sodium amide ($NaNH_2$) can react with these protons.

This reaction gives an **organometallic** compound which has both organic and metallic parts. This alkynide is also commonly called an acetylide. These carbanions are good nucleophilic reagents which react with electrophiles. If an alkyl halide is used as the electrophile, it gives a very useful C–C bond-forming process as shown in Figure 7.12. This second step is an example of the S_N2 nucleophilic substitution reaction shown in Figure 7.16.

$$R-C\equiv C-H \xrightarrow{\text{NaNH}_2} R-C\equiv C^{\ominus}Na^{\oplus} \xrightarrow{\text{R'Br}} R-C\equiv C-R'$$

Terminal alkyne   Alkynide carbanion   Extended alkyne

FIGURE 7.12
Chain building alkynide reaction sequence.

7.5 ALKYL HALIDES

As shown in Figure 7.1, the highly polarized carbon–halogen bond makes alkyl halides key compounds in functional group interconversions. This is because alkyl halides can take part in several different reactions. The polarity of the carbon–halogen bond leads to heterolytic replacement of the halide by a new bond. This is usually done by nucleophilic substitution and elimination reactions. Both of these reaction types can occur on the same alkyl halide substrates. This is an example of competing reactions.

7.5.1 Carbon–Halogen Bond Polarity

The polarization of the carbon–halogen bond lets alkyl halides react with silver nitrate, $AgNO_3$, to give precipitates of silver halides. This is shown in Figure 7.13, and is a simple visual test for alkyl halides.

$$R-X \xrightarrow{\text{AgNO}_3/\text{H}_2\text{O}} R-OH + AgX_{(s)}$$

FIGURE 7.13
Visual test for alkyl halides.

The formation of Grignard reagents is another important metal-based reaction which relies on this bond polarity. As Figure 7.14 shows, magnesium is placed between carbon and halide. This gives another organometallic reagent which is a useful carbanion. These carbanions react as nucleophiles in reactions to give new C–C bonds.

$$R-X \xrightarrow[\text{ether}]{\text{Mg}} \overset{\delta^{\ominus}}{R}-\overset{\delta^{\oplus}}{MgX}$$

FIGURE 7.14
Formation of Grignard reagents.

7.5.2 Nucleophilic Substitution

Nucleophilic substitution is one of the major functional group reactions. It is simply the replacement of a leaving-group ligand by an incoming nucleophile ligand. There is no change in the nominal oxidation number at the carbon center of interest. There is also no change in the bond order. Compare this with addition, Section 7.3.1, and elimination, Section 7.5.3. The reactions are shown by the equations in Figure 7.15. The only difference between these two examples is whether or not the nucleophile has a formal negative charge.

As shown in Table 7.2, there are many different possible nucleophiles. Therefore, there are many different products from nucleophilic substitution.

Table 7.2   Selected Alkyl Halide R–X Nucleophilic Substitutions

Nucleophile		Product	Class
Hydroxide	$HO^{\ominus}$	HOR	Alcohol
Alkoxide	$R^1O^{\ominus}$	R^1OR	Ether
Thioalkoxide	$R^1S^{\ominus}$	R^1SR	Thioether
Cyanide	$N\equiv C^{\ominus}$	$N\equiv CR$	Nitrile
Alkynide	$R^1C\equiv C^{\ominus}$	$R^1C\equiv CR$	Alkynes
Carboxylate	$R^1CO_2^{\ominus}$	R^1CO_2R	Ester
Ammonia	$H_3N\colon$	$\overset{\oplus}{R}NH_3\ \overset{\ominus}{X}$	Ammonium halide

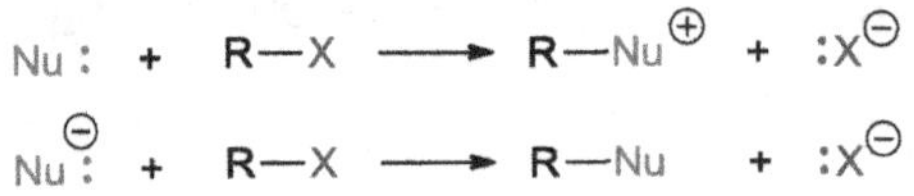

$$Nu\colon \;+\; R\!-\!X \longrightarrow R\!-\!\overset{\oplus}{Nu} \;+\; \colon\!\overset{\ominus}{X}$$

$$\overset{\ominus}{Nu}\colon \;+\; R\!-\!X \longrightarrow R\!-\!Nu \;+\; \colon\!\overset{\ominus}{X}$$

FIGURE 7.15
General nucleophilic substitution.

A nucleophilic substitution follows two main mechanisms, depending on the substrate, nucleophile, and solvent. As Figure 7.13 shows, these are labeled as S_N1 and S_N2 to show which pathway has been followed.

The difference between these mechanisms is the role of the nucleophile. In an S_N2 process, the reaction is started by the attack of the nucleophile at the polarized electrophilic carbon. This concerted process goes through a transition state in which partial bond forming and breaking occurs. Then, this five-centered state breaks down to give the final substituted product.

The rate of reaction depends on the concentration of both substrate and nucleophile. In other words, it shows a **bimolecular** dependence on two different species. Therefore, this mechanism is described as S_N2 (substitution nucleophilic bimolecular).

The alternative S_N1 process is stepwise. The first step is the heterolysis to give an intermediate carbocation. This carbocation intermediate reacts with the nucleophile to complete the substitution. The reaction rate depends only on the concentration of the substrate and not on the nucleophile. In other words, it shows a **unimolecular** dependence on a single species. Therefore, this mechanism is described as S_N1 (substitution nucleophilic unimolecular).

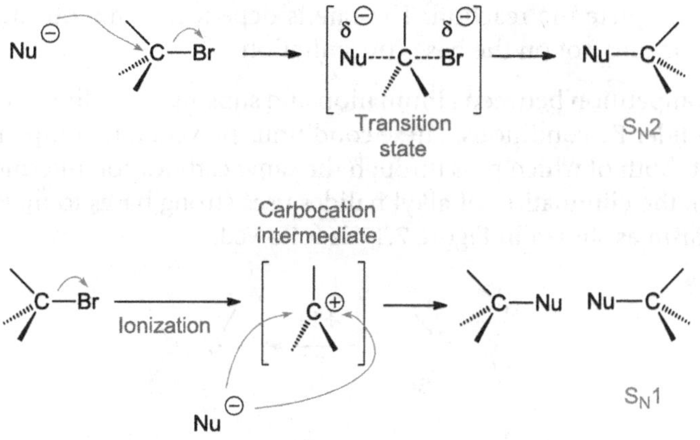

FIGURE 7.16
Summary mechanisms of nucleophilic substitution.

There are stereochemical outcomes for the above mechanisms. In the S_N2 process, an inversion of the configuration occurs during the movement from sp^3-tetrahedral→trigonal bipyramidal→sp^3-tetrahedral.

The S_N1 process passes through a carbocation. In Chapter 5 we saw that an sp^2-carbocation was planar. Because of this, the nucleophile can approach from either side of the plane. This gives a mixture of products from either inversion or retention of the configuration.

Appendix 9 gives more information about the stereochemistry of nucleophilic substitution reactions.

7.5.3 Elimination Reactions

As mentioned earlier, alkyl halides can undergo elimination reactions in competition with substitution. Elimination causes an increase in the bond order, and is the reverse of the electrophilic addition reactions of alkenes and alkynes seen in Section 7.3.1. The highly polarized carbon–halogen bond is again important in these reactions.

As with substitution, there are two main mechanisms for elimination. The elimination equivalents for nucleophilic substitution are:

- E2 for S_N2;
- E1 for S_N1.

These mechanisms show the same rate-determining concentration dependencies as for the substitutions.

Elimination by E2 is a concerted process. A base removes a proton from the carbon next to the halide attachment, and the halide is lost at the same time. Therefore, the rate depends on both the substrate and base concentration.

Elimination by E1 is stepwise. A carbocation is formed by heterolysis of the carbon–halogen bond. Then, the carbocation loses a proton from a neighboring carbon to complete the reaction. The rate is dependent only on the substrate concentration and not on the base concentration.

To avoid competition between elimination and substitution, elimination is usually done under E2 conditions. These conditions prevent any competition from E1 and S_N1, both of which pass through the same carbocation intermediate. For this reason, the elimination of alkyl halides uses strong bases to make sure the E2 mechanism as shown in Figure 7.17 is followed.

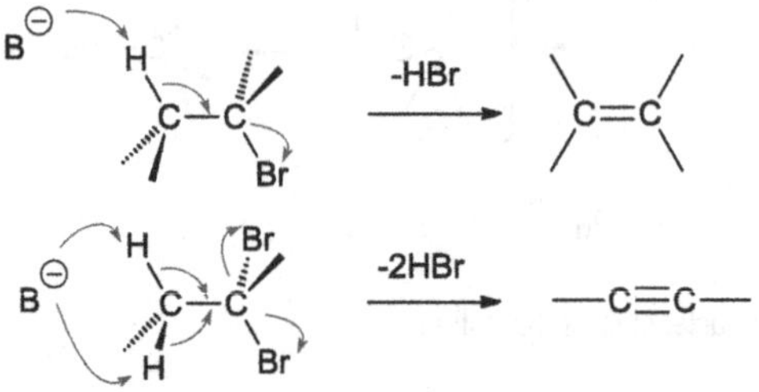

FIGURE 7.17
Examples of E2 elimination reactions.

7.6   ALCOHOLS AND ETHERS

The alcohol functional group has different bonds which can react in various ways. In Chapter 6 we discussed the acidity of the O–H bond. The elimination of the components of water is called **dehydration**. If this elimination occurs between adjacent carbon atoms, it gives an alkene. The replacement of the OH group gives substitution. Finally, removal of H_2 across the alcohol C–O bond is oxidation and gives a carbonyl C=O product.

7.6.1   Making and Using Alkoxides

As discussed in Chapter 6, alkoxides are formed by the deprotonation of alcohols. Figure 7.18 shows one way to do this by reaction of an alkali metal, usually Na, with an alcohol. This reaction gives off H_2 and is a visual test for the presence of an OH group.

FIGURE 7.18
Visual test for alcohols, alkoxide formation.

As Figure 7.19 shows, these oxyanions can have different purposes. They are often used as bases for the elimination reactions in Section 7.5.3. Alternatively, they can be nucleophiles for the substitutions in Section 7.5.2.

In Figure 7.19, the substitution reaction shown is an example of a general reaction which is called Williamson ether synthesis. In this reaction, an alkoxide nucleophile is used to give either symmetrical or unsymmetrical ethers.

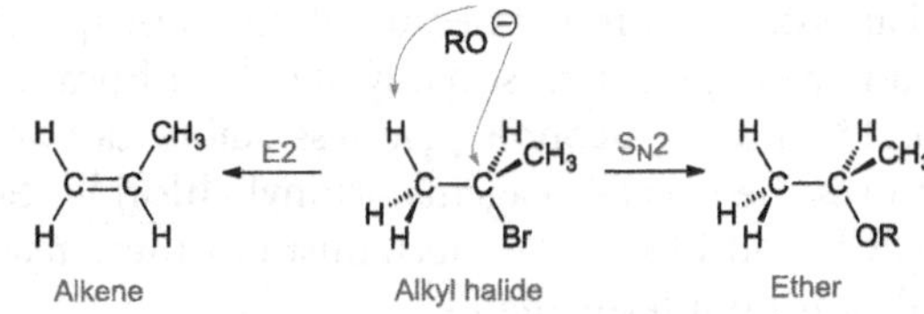

FIGURE 7.19
Selected applications of metal alkoxides.

7.6.2   Alcohol Dehydration

Alcohols can lose the elements of water to give alkenes. However, the hydroxide group is a good base and nucleophile. Therefore, it is a poor leaving group. Because of this, elimination is done under acidic conditions. The acid acts as a catalyst to protonate the alcohol –OH group. This gives a protonated alkyloxonium ion intermediate. The leaving group is now a neutral water molecule. Figure 7.20 shows this sequence for this E1 reaction.

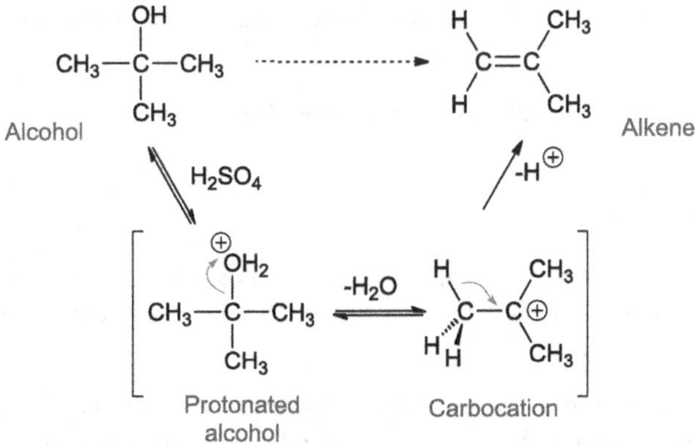

FIGURE 7.20
Acid-catalyzed alcohol dehydration.

7.6.3   Alcohol Substitution

Figure 7.21 shows one of the most useful examples of alcohol substitution. This reaction is used in the preparation of alkyl halides. This reaction is the reverse of the nucleophilic substitution of halide by hydroxide, which we saw in Section 7.5.2. The poor leaving group property of –OH is again improved to allow reaction to occur.

One way to improve the leaving group is to react the alcohol with concentrated hydrohalous acids (HCl, HBr, HI). As with elimination in Figure 7.20, protonation of the alcohol oxygen gives an alkyloxonium ion. Loss of a neutral water molecule gives the carbocation. This reacts with halide ions to give the alkyl halide product. This process follows an S_N1 mechanism and needs a relatively stable carbocation.

This acid-catalyzed approach is limited because it does not apply to all alcohols. Tertiary alcohols can easily give a reasonably stable carbocation. However, as seen in Chapter 5, 1° and 2° alcohols give less stable carbocations. A more general alternative uses the special reagents thionyl chloride ($SOCl_2$) or phosphorus trihalides (PCl_3 and PBr_3). The mechanism of these reactions does not go through free carbocation intermediates.

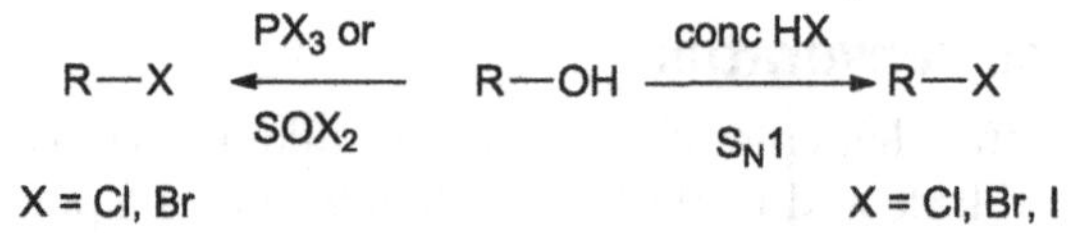

$$R{-}X \xleftarrow[SOX_2]{PX_3 \text{ or}} R{-}OH \xrightarrow[S_N1]{conc\ HX} R{-}X$$

$$X = Cl, Br \qquad\qquad X = Cl, Br, I$$

FIGURE 7.21
Alkyl halides from alcohols.

7.6.4 Alcohol Oxidation

The oxidation chemistry of alcohols depends on whether the substitution at alcohol carbon is 1°, 2°, or 3°. As Figure 7.22 shows, alcohols can be com-

$$CH_3CH_2OH + 3O_2 \longrightarrow 2CO_2 + 3H_2O$$

FIGURE 7.22
Combustion of alcohols.

busted in oxygen to give CO_2 and H_2O. This is similar to the combustion of alkanes in Section 7.2.1, but the heat given off is less because one carbon is already in an oxidized state.

Many oxidation reagents are available. However, we can see the principles in the dichromate system of $Na_2Cr_2O_7$ or $K_2Cr_2O_7$ in aqueous acid. As Figure 7.23 shows, the number of hydrogen ligands on the alcohol carbon determines how much oxidation can occur.

Oxidation simply means the removal of H_2 from the C–OH bond. Therefore, 1° alcohols give aldehydes, and 2° alcohols give ketones. Because aldehydes still have a hydrogen on the carbonyl carbon, they can be further oxidized to carboxylic acids. With 3° alcohols, no simple oxidation can occur because they have no hydrogen on the alcohol carbon.

7.6.5 Ether Cleavage

Ethers are relatively unreactive derivatives of alcohols. Because of this, several ethers are widely used as polar solvents. As seen in Figure 7.24, strong acid is reacted with the oxygen Lewis base in order to break the O–C bond in an ether. The protonation of the oxygen gives a dialkyloxonium ion. This ionizes by S_N1, or is subject to S_N2 attack by the conjugate base nucleophile to give cleavage products.

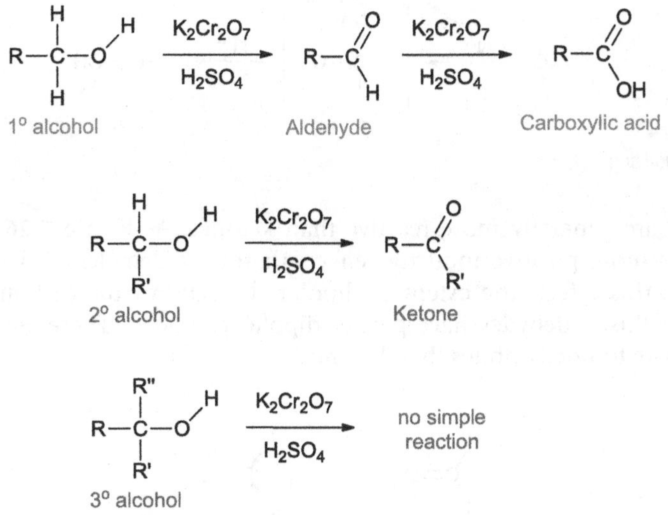

FIGURE 7.23
Oxidation of 1°, 2°, and 3° alcohols.

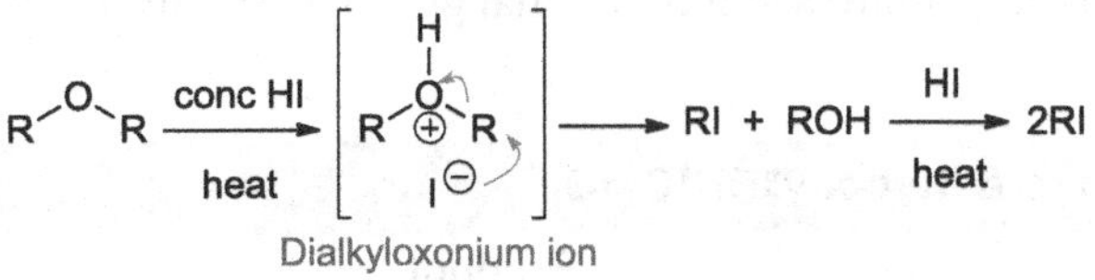

FIGURE 7.24
Ether cleavage.

7.7 ALDEHYDES AND KETONES

In Chapter 2, the dipolar nature of the carbonyl group was shown as an electrophilic carbon and a nucleophilic oxygen. Because the carbonyl group has a double bond, addition is a major reaction type.

In Chapter 6, the carbonyl group was shown to increase the acidity of any hydrogens on the α-carbon atom. This allows the formation of a resonance-stabilized **enolate** carbanion, and gives a second pathway for the reaction of aldehydes and ketones.

7.7.1 Nucleophilic Addition

The carbonyl dipole controls the initial reaction site. As Figure 7.25 shows, the reaction usually starts with nucleophilic attack at the electrophilic carbonyl carbon. Therefore, the process is called nucleophilic addition. Compare this with the electrophilic addition reactions of alkenes in which the initial attack was by the C=C on an electrophile.

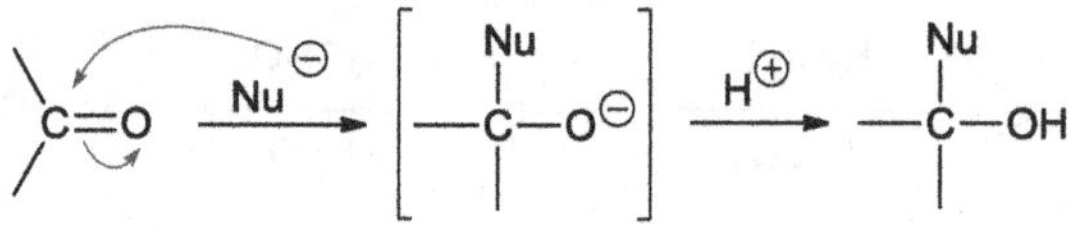

FIGURE 7.25
Nucleophilic addition.

Aldehydes are generally more reactive than ketones. As Figure 7.26 shows, a ketone has more positive inductive effects than an aldehyde. As discussed in Chapter 2, this affects the extent of dipolar character of the carbonyl group. Because of this, aldehydes have greater dipolar character. Therefore, they are more reactive to nucleophiles than ketones.

FIGURE 7.26
The relative inductive effects in aldehydes and ketones.

Table 7.3 lists different nucleophilic addition reactions to carbonyl compounds. These are a very important set of functional group interconversions.

Table 7.3	Nucleophilic Addition to a Carbonyl $RR^1C{=}O$		
Addition Reagent		**Product**	**Class**
Hydrogen cyanide	HCN	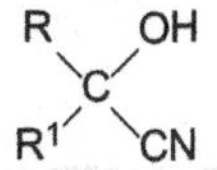 R, OH, C, R^1, CN	Cyanohydrin
Water	H_2O	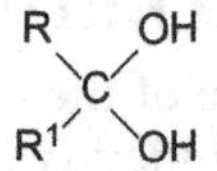 R, OH, C, R^1, OH	Hydrate
Alcohol	R^2OH	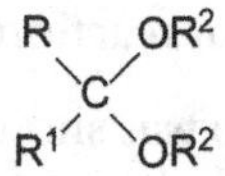 R, OR^2, C, R^1, OR^2	Acetal
Hydrogen	$H^{\ominus}/H^{\oplus}$	R, OH, C, R^1, H	Alcohol
Grignard	R^2MgX	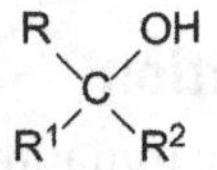 R, OH, C, R^1, R^2	Alcohol
Amine	R^2NH_2	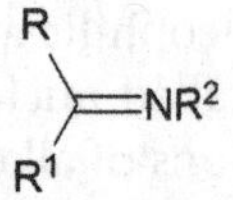 $R, {=}NR^2, R^1$	Imine

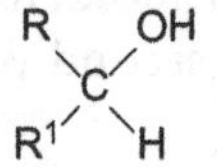

7.7.1.1   *CYANOHYDRIN FORMATION*

The first step of HCl addition is attack of the nucleophilic cyanide. As Figure 7.27 shows, this addition gives the product cyanohydrin with one extra carbon in the chain. The cyanohydrin can be readily changed to other useful products. For example, hydrolysis of the nitrile gives α-hydroxy acids.

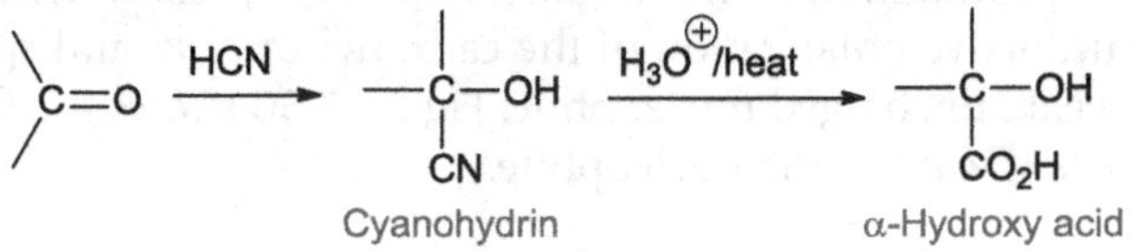

FIGURE 7.27
Cyanohydrin formation and hydrolysis sequence.

7.7.1.2   *GRIGNARD ORGANOMETALLIC ADDITION*

In Section 7.5.1, we showed the formation of Grignard reagents. In Figure 7.28, we see that the addition of these carbanions to carbonyl compounds gives an intermediate magnesium alkoxide. Protonation of the alkoxide gives the alcohol product. Aldehydes give 2° alcohols and ketones give 3° alcohols.

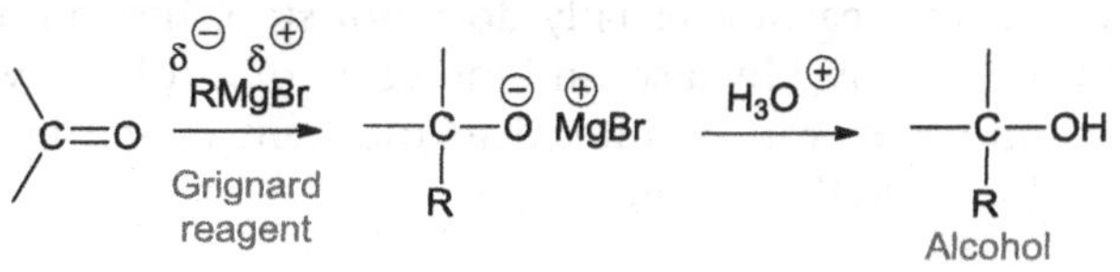

FIGURE 7.28
Grignard addition sequence.

7.7.1.3   *HYDRIDE ADDITION*

Hydride addition is a chemical reduction. It uses metal hydride reagents. The most common of these are lithium aluminum hydride ($LiAlH_4$) and sodium borohydride ($NaBH_4$).

In this addition, the nucleophile is the reactive hydride ion, $H^{\ominus}$. The second hydrogen is by protonation of the intermediate alkoxide. The overall reaction is a reduction. As Figure 7.29 shows, this changes carbonyl compounds into alcohols. Aldehydes give 1° alcohols and ketones give 2° alcohols.

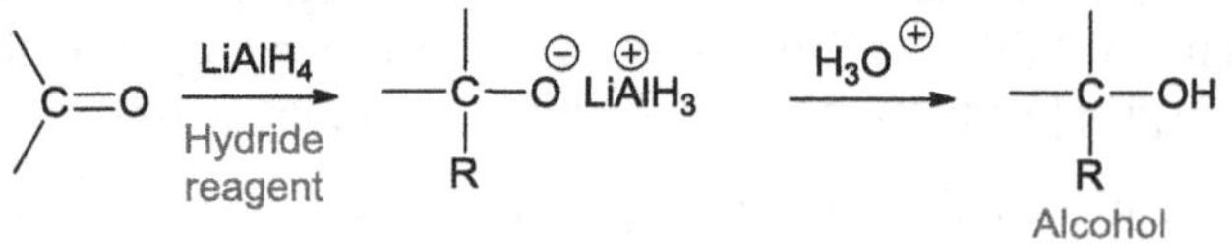

FIGURE 7.29
Hydride reduction of the carbonyl group.

7.7.1.4 ADDITION OF WATER AND ALCOHOLS

The nucleophiles in these reactions are weaker non-ionic species. The relative stability of substrate and product determines the position of the equilibrium in these reactions. To shift the equilibrium toward product, the rate of the forward process must be increased. This is usually done by acid catalysis.

The first step is protonation of the negatively polarized carbonyl oxygen. This increases the inductive polarization of the carbonyl carbon, and speeds up the attack by the weak, uncharged nucleophile. Figure 7.30 shows the formation of hydrate products if water is the nucleophile.

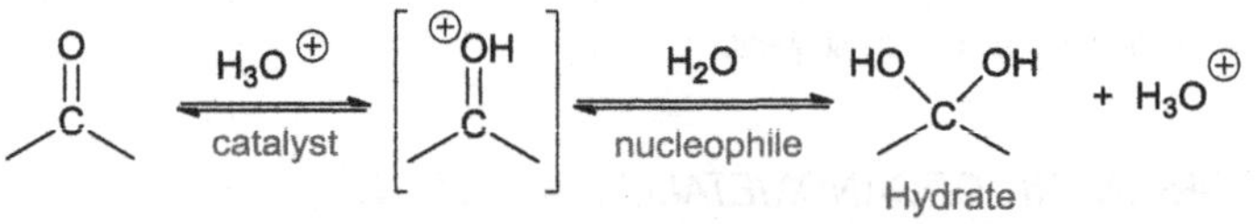

FIGURE 7.30
Hydration of carbonyl compounds.

As Figure 7.31 shows, if an alcohol is used as the nucleophile, a hemiacetal is made. In this case, the reaction usually does not stop. Instead, nucleophilic substitution of the hydroxyl by a second molecule of alcohol gives an acetal. The reaction equilibrium can be moved to the right or left by addition or removal of the H_2O or ROH reagent.

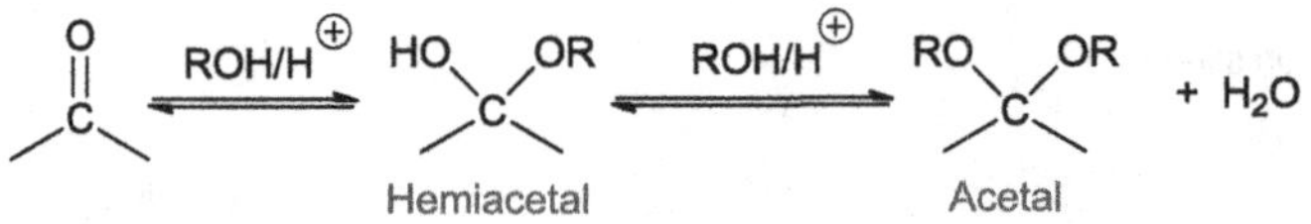

FIGURE 7.31
Acetal formation from carbonyl compounds.

The hemiacetal and acetal functional groups are common in cyclic simple carbohydrates and in complex carbohydrates. Chapter 8 discusses examples of these compounds.

7.7.2 Condensation Reactions with 1° Amino Derivatives

A **condensation** reaction is made up of an addition and elimination reaction sequence. As Figure 7.32 shows, ammonia and 1° amine derivatives can be used in these reactions. Table 7.4 lists some condensation examples with aldehydes and ketones to give products in which the nitrogen replaces the carbonyl oxygen.

These products are usually crystalline solids with sharp melting points. This feature means these **derivative** products can be used as standards to identify many common aldehydes and ketones.

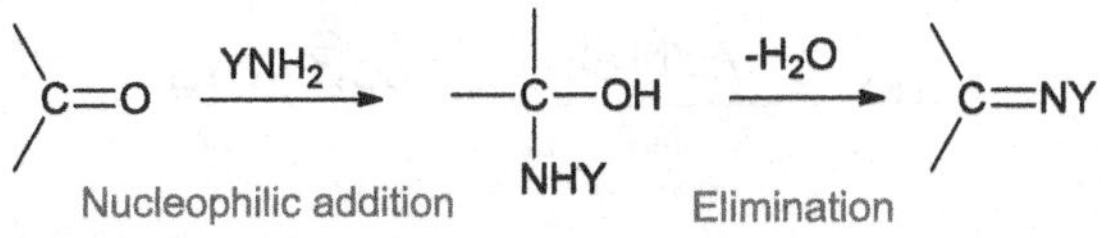

FIGURE 7.32
Primary amino-carbonyl condensation reaction.

Table 7.4	**Amino-Carbonyl Condensation Products**		
Amino	**Reagent**	**Product**	**Class**
Amine	RNH_2	$=NR$	Imine
Hydroxylamine	H_2NOH	$=NOH$	Oxime
Hydrazine	$RNHNH_2$	$=NNHR$	Hydrazone

The best known example is the 2,4-dinitrophenylhydrazone derivative, or 2,4-DNP. Figure 7.33 shows how these are made by reaction of carbonyl compounds with Brady's hydrazine reagent to give the orange-red colored derivatives. The appearance of the 2,4-DNP precipitate is a visual test for aldehydes and ketones.

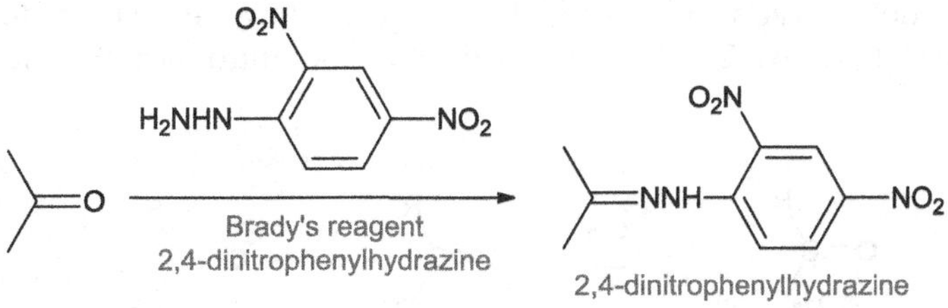

FIGURE 7.33
Brady's test for aldehydes and ketones.

7.7.3 Oxidation of Aldehydes

In Section 7.6.4 we saw the oxidation of aldehydes to carboxylic acids as part of the oxidation of 1° alcohols. In addition to preparing carboxylic acids, this reaction is a simple method to tell whether an aldehyde or a ketone is present.

In a reaction known as the Tollen's silver mirror test, oxidation is done with an aqueous ammonia/silver solution. During the reaction shown in Figure 7.34, silver cations are reduced to metallic silver. The silver metal coats the reaction vessel with a shiny mirror. This is a positive visual test for an aldehyde.

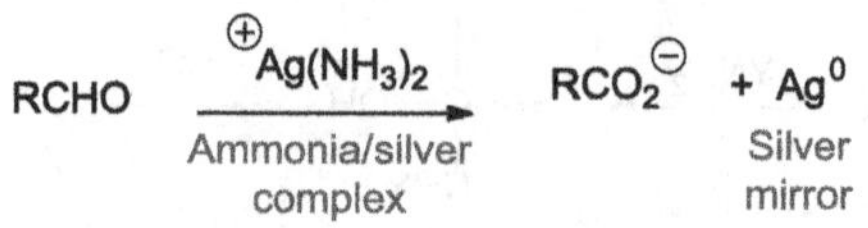

$$RCHO \xrightarrow[\text{Ammonia/silver complex}]{\overset{\oplus}{Ag(NH_3)_2}} RCO_2^{\ominus} + Ag^0$$

Silver mirror

FIGURE 7.34
Tollen's silver mirror test for aldehydes.

7.7.4 Reaction at the Carbonyl α-Carbon

The chemistry of acidic hydrogens on the α-carbons of carbonyl compounds is extensive. In this book, we show the concept by taking the example of the haloform reaction in Figure 7.35. This reaction is not only a preparation of carboxylic acids, but is also a visual test for the $-CO \cdot CH_3$ acetyl group.

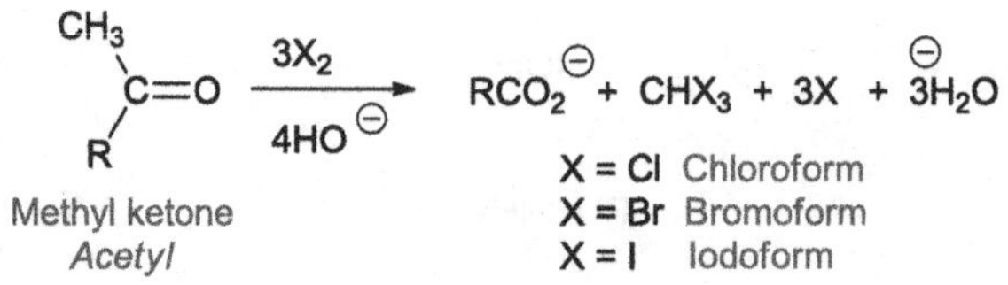

$$\underset{\substack{\text{Methyl ketone}\\\textit{Acetyl}}}{\overset{CH_3}{\underset{R}{C}}{=}O} \xrightarrow[4HO^{\ominus}]{3X_2} RCO_2^{\ominus} + CHX_3 + 3X^{\ominus} + 3H_2O$$

X = Cl Chloroform
X = Br Bromoform
X = I   Iodoform

FIGURE 7.35
The haloform reaction.

Although the reaction is general, it is usually done with iodine because it gives iodoform as solid yellow precipitate. Figure 7.36 shows the mechanism of this reaction and the principles of enolate chemistry.

In the main reaction, hydroxide deprotonates the α-carbon to give an enolate. Then, the enolate reacts with iodine. This is repeated two more times to give the triiodoacetyl derivative. The final step in the sequence introduces the chemistry of

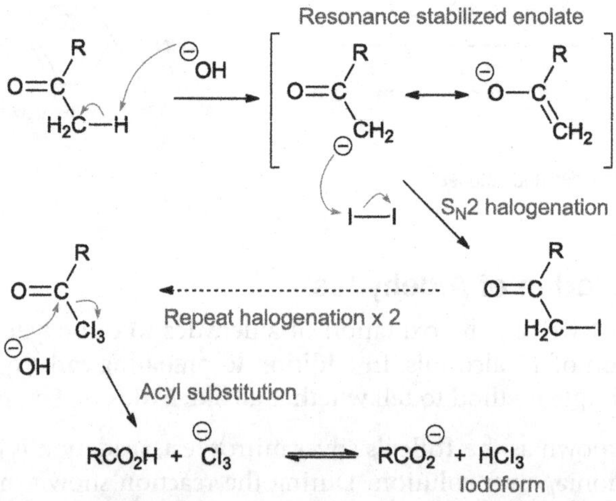

FIGURE 7.36
Mechanism of the iodoform reaction.

acyl derivatives in Section 7.8. In this, hydroxide substitutes the triiodocarbanion to give the carboxylic acid. After proton exchange, the solid iodoform is deposited.

7.8 CARBOXYLIC ACIDS AND ACYL DERIVATIVES

Chapter 2 gave the structural features of carboxylic acids and acyl derivatives. These compounds can be seen as functional classes with one heteroatom bond which is modified by a carbonyl group. In Chapter 6, we saw how this explains the acidity of carboxylic acids in which the alcohol and carbonyl parts acted together. The acidic nature of carboxylic acids was limited to the O–H bond.

In Section 7.7, we saw how the reactions of the carbonyl group were mostly started by nucleophilic attack to give addition products. Nucleophilic attack is also the major reaction of acyl derivatives. However, instead of overall addition, the reaction leads to substitution products as shown in Figure 7.37.

7.8.1 Nucleophilic Acyl Substitution

The change from addition to substitution is because of the potential leaving group Y in Figure 7.37. In aldehydes and ketones, there is no simple leaving group. Only hydride or carbanions are possible. Because these are both very strong nucleophiles, they are very poor leaving groups. Acyl derivatives such as acyl halides, acid anhydrides, esters, carboxylic acids, and amides have better leaving groups. These include halide, carboxylate, alkoxide, hydroxide, and amine anion.

The overall substitution reactions give the same result as the one-step S_N2 reactions in Section 7.5.2. However, they occur by a different mechanism. Figure 7.37 shows nucleophilic acyl substitution as a two-step process which goes through an intermediate oxyanion.

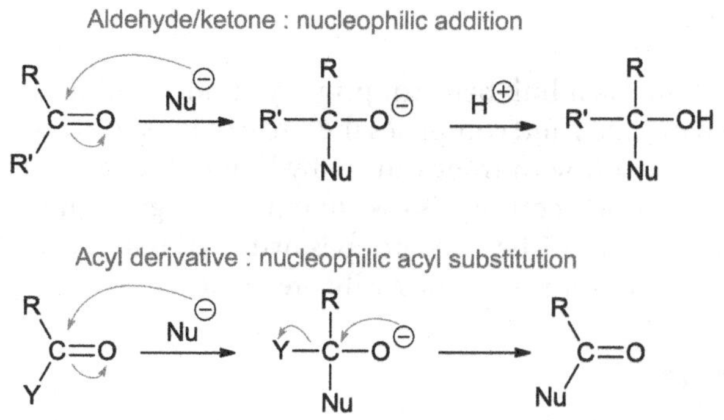

FIGURE 7.37
Nucleophilic addition versus acyl substitution.

The relative reactivity of acyl derivatives in substitution is controlled by the polarization in each derivative. The higher the polarity of the C–Y bond, the more reactive the acyl derivative. Figure 7.38 lists the acyl derivative reactivity as acyl halides > acid anhydrides > esters > amides.

All the acyl derivatives can be prepared from the carboxylic acid. Because direct conversion needs the substitution of a hydroxyl group, these are done under acidic conditions to improve the leaving group. This method of using acid catalysis is the same as we saw for alcohols in Section 7.6.3 and for hydrates in Section 7.7.1.4.

Alternatively, all acyl derivatives can be made from the very reactive acyl halides. These most reactive members are prepared from the acid with the reagents PCl_3 or $SOCl_2$. Any acyl derivative can be prepared by nucleophilic acyl substitution of an acyl derivative of higher reactivity.

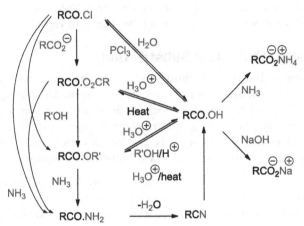

FIGURE 7.38
Acyl substitution reactivity order.

Figure 7.38 also shows a link between primary amides and nitriles. This link is the elimination of water, and completes the circle of interconversions. In Section 7.7.1.1 it was shown how nitriles can be hydrolyzed to give carboxylic acids. This is a useful reaction because it is easy to put a –CN group into a molecule by nucleophilic substitution. The –CN group is hydrolyzed to give a $-CO_2H$ group. From this, all acyl derivatives can be easily prepared.

7.8.2 Esters

Esters are an important class of compounds. Because of their importance, we will look at the mechanism of their preparation. Figure 7.39 shows one way to make esters by the acid-catalyzed reaction of esterification. This uses a carboxylic acid and an alcohol, along with an acid catalyst. During this reaction a molecule of water is formed.

FIGURE 7.39
Preparation of esters.

This equation shows the overall forward reaction. However, the acid-catalyzed process is actually an equilibrium. As Figure 7.40 shows, this equilibrium can be shifted by having different reagents in excess.

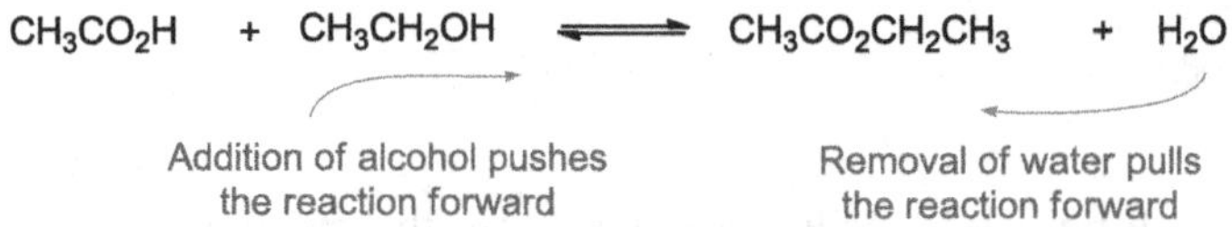

FIGURE 7.40
The reversibility of esterification.

To understand this, you need to study the mechanisms of esterification and reverse hydrolysis in Figure 7.41. This clearly shows there are several equilibrium steps in the overall process.

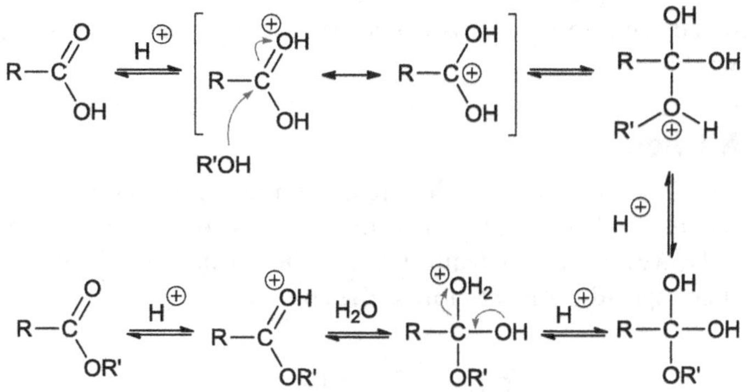

FIGURE 7.41
Mechanism of acid-catalyzed esterification.

The hydrolysis of an ester to give a carboxylic acid and an alcohol can also be done under basic conditions. This alternative reaction is by a standard nucleophilic acyl substitution.

7.8.3 Amides

As shown in Figure 7.38, amides can be formed by substitution reactions of the acyl derivatives above them in the reactivity chart. Chapter 2 showed how amides are classified as 1°, 2°, and 3° amides depending on the substitution at the nitrogen. The amide bond is important in many natural products of biological importance, such as peptides and proteins. Figure 7.42 shows why the bonding in an amide makes it special.

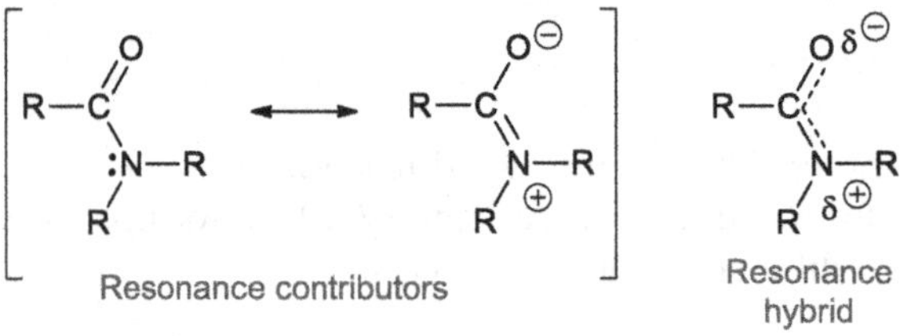

FIGURE 7.42
The amide bond.

The nitrogen lone pair is easily delocalized into the carbonyl group. The result of delocalization is best drawn as a resonance hybrid. This gives partial C=N character to the amide bond, and a planar sp^2-hybridized state to the nitrogen. This stops free rotation around the C–N bond and can cause possible *cis/trans* isomers.

Proof of this limited rotation is seen in the fixed conformations in amides. Amides also have a shorter C–N bond length than amines, ±132 pm compared with ±147 pm. This delocalization in amides also explains their non-basic character compared with amines. This is because the lone pair is no longer free to act as a Lewis base.

7.8.4   Nitriles

In Section 7.5.2, we saw that nitriles are easily made and can be hydrolyzed to carboxylic acids. See Figure 7.43. In Section 7.8.5, you can see that a nitrile can be reduced. However, the key feature is that the formation of a nitrile and any subsequent change adds one carbon to the chain.

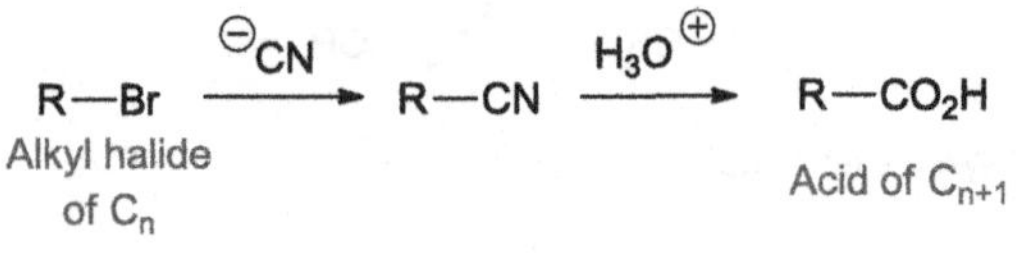

FIGURE 7.43
Nitrile chain lengthening sequence.

7.8.5   Reduction

As you can see in Figure 7.44, reduction is a major reaction of all acyl derivatives. This can be done with metal hydride reagents, such as $LiAlH_4$. With acyl halides, anhydrides, acids, and esters, reduction gives alcohols as the products. For amides and nitriles, reduction gives amines as the products.

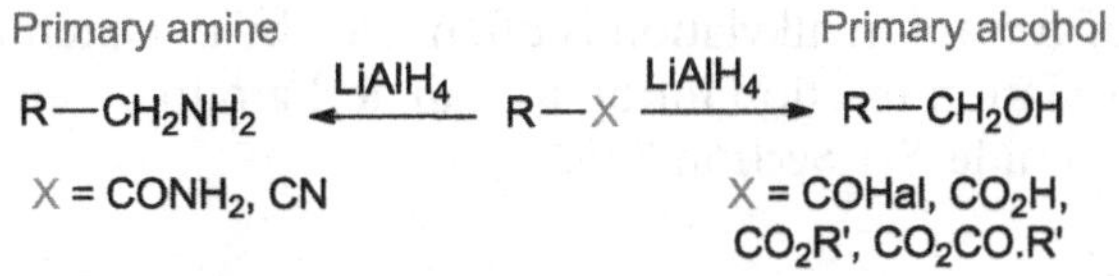

Primary amine

$$R-CH_2NH_2 \xleftarrow{\text{LiAlH}_4} R-X \xrightarrow{\text{LiAlH}_4} R-CH_2OH$$

Primary alcohol

$X = CONH_2, CN$

$X = COHal, CO_2H,$
$CO_2R', CO_2CO.R'$

FIGURE 7.44
Hydride reduction of acyl derivatives.

7.9   AMINES

Chapter 6 examined the structure and Lewis basicity of amines. The chemistry of amines is mostly about the nitrogen lone pair and its nucleophilic or basic properties. Therefore, amine reactions are based on either nucleophilic substitution or acid–base types. Common substrates for nucleophilic substitution are alkyl and acyl halides. Reaction of these substrates gives alkylation and acylation products.

7.9.1   Alkylation

Amines are derivatives of ammonia. They are prepared by replacement of the hydrogens of NH_3 with carbon groups. Figure 7.45 shows ammonia and amines reacting as non-ionic nucleophiles in S_N2 type reactions with alkyl halides. These **alkylation** reactions add alkyl groups and give higher order amines.

Positive inductive effects from alkyl groups cause an amine to become more nucleophilic, as it has more alkyl groups attached to it. In fact, it can be very difficult to stop the process at any point during alkylation. Because of this lack of control, mixtures of products are usually formed.

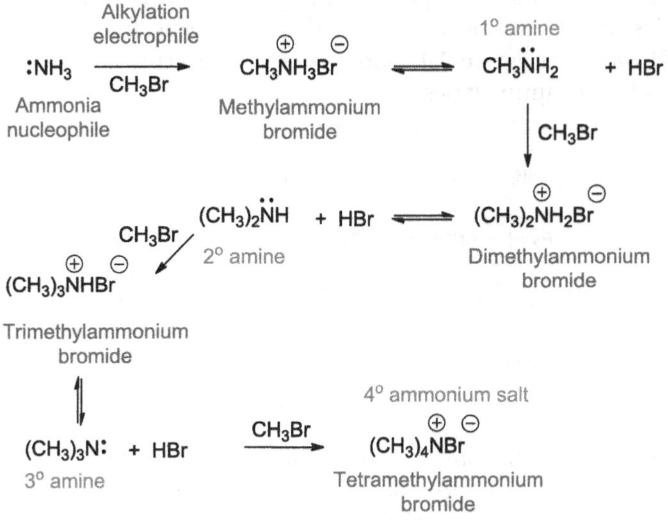

FIGURE 7.45
Stepwise amine alkylation.

This problem of control in alkylation is often fixed by the alternative **acylation** approach. An acylation reaction introduces an acyl group on an amine nucleophile to give an amide. See Section 7.9.2.

7.9.2 Acylation

Section 7.8.3 showed how amines react with acyl derivatives to give amides. Figure 7.46 shows that 1° and 2° amines can react to give amides by substitution. Because 3° amines have no hydrogen substituent to lose, they cannot give amides.

In Chapter 6, we saw that the nitrogen in amides is not nucleophilic. Because of this, the acylation reaction stops at monosubstitution product. Then, the amides can be reduced with $LiAlH_4$ to amines. This solves the problem of overalkylation as shown in Figure 7.45.

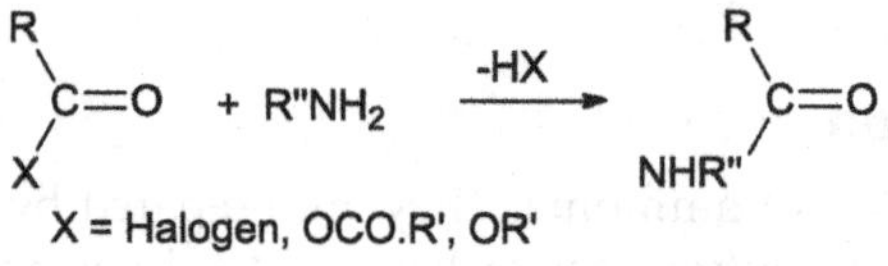

$$\underset{X}{\overset{R}{\diagdown}}C=O \ + \ R''NH_2 \ \xrightarrow{\ -HX\ } \ \underset{NHR''}{\overset{R}{\diagdown}}C=O$$

X = Halogen, OCO.R', OR'

FIGURE 7.46
Amine acylation.

7.9.3 Diazotization

Diazotization is an important reaction of 1° amines. In the diazotization process, the NH_2 group is changed to a diazonium salt, $R-N_2{}^+X^-$. This is done by reaction with nitrous acid (HNO_2). The reactive salt is not usually isolated. Loss of a gaseous N_2 molecule gives a carbocation which can react with various nucleophiles. We do not need to study these reactions and their mechanisms in detail. However, as Figure 7.47 shows, the nitrous acid reaction gives a visual method to identify amine types.

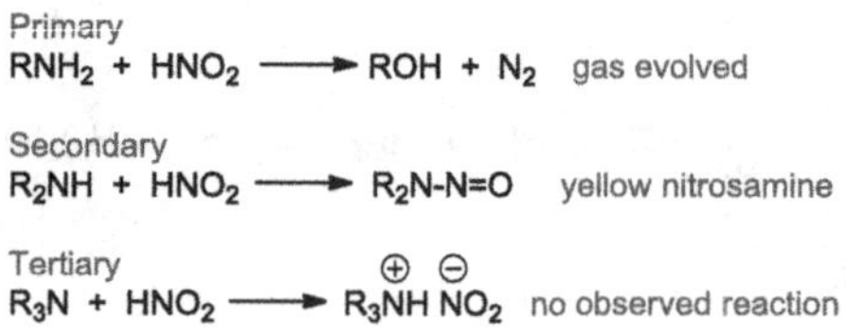

Primary
$$RNH_2 \ + \ HNO_2 \ \longrightarrow \ ROH \ + \ N_2 \quad \text{gas evolved}$$

Secondary
$$R_2NH \ + \ HNO_2 \ \longrightarrow \ R_2N\text{-}N=O \quad \text{yellow nitrosamine}$$

Tertiary
$$R_3N \ + \ HNO_2 \ \longrightarrow \ R_3\overset{\oplus}{N}H \ \overset{\ominus}{NO_2} \quad \text{no observed reaction}$$

FIGURE 7.47
Amine diazotization classification.

Primary amines give diazonium salts. When these lose N_2, the carbocations react with water to give alcohols. Secondary amines only have one hydrogen on the nitrogen. They cannot complete the diazotization reaction and give yellow oily nitrosamine products. Finally, 3° amines, with no hydrogens on nitrogen, simply undergo an acid–base reaction with nitrous acid. This gives soluble salts. As a result, there is no visible reaction.

7.10 AROMATIC COMPOUNDS

We have talked about aromatic compounds simply as structural examples of the functional classes. Details of aromatic compounds are limited to the simplest member, benzene, and its derivatives. In general, you saw in Chapter 2 how the aromatic system Ar– is an alternative for the alkyl R–.

Most functional group reactions of compounds with an aromatic substituent follow those discussed in earlier sections. However, in Chapter 4 we saw that arenes have a special delocalized system which gives large resonance effects. Because of this, the reactivity of functional groups which are attached directly to the aromatic ring may change. We have already seen the example of the increased acidity of phenols in Chapter 6.

7.10.1 Electrophilic Aromatic Substitution

As with alkenes, the benzene system is electron rich and can act as a nucleophile. Because of delocalization of the π-electrons, this nucleophile is weak and has lower reactivity with electrophiles. Very reactive electrophiles and/or Lewis acid catalysts are used to promote the reaction.

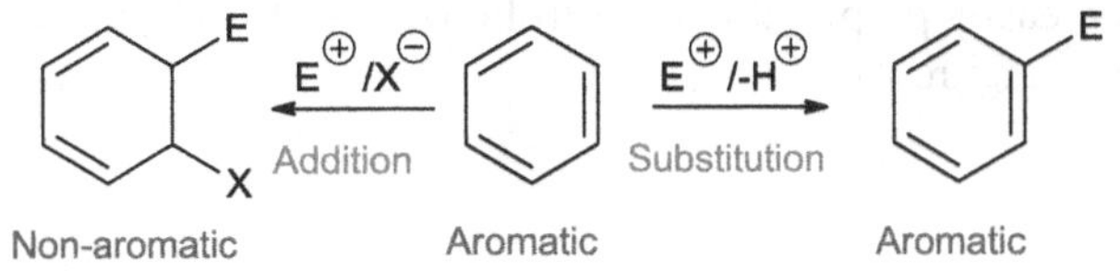

FIGURE 7.48
Aromatic electrophilic addition versus substitution.

However, alkenes react by electrophilic addition, but the benzene ring reacts by electrophilic substitution. This is because the benzene nucleus wants to keep the extra aromatic resonance stabilization. As Figure 7.48 shows, this benefit is lost in an addition reaction because the delocalized system is broken.

The reaction follows a stepwise mechanism. The rate-determining addition of the electrophile is the first step. Then, a proton is eliminated to complete the aromatic system again. A wide choice of electrophiles allows many different substituted aromatic compounds to be prepared in this way. Some of most important ones are summarized in Table 7.5.

The position of electrophilic substitution when substituents are already on the benzene ring can be difficult to predict. The position of substitution depends on the relative position and type of the substituents which are already present. However, this is beyond the scope of this book.

Table 7.5   Aromatic Electrophilic Substitution

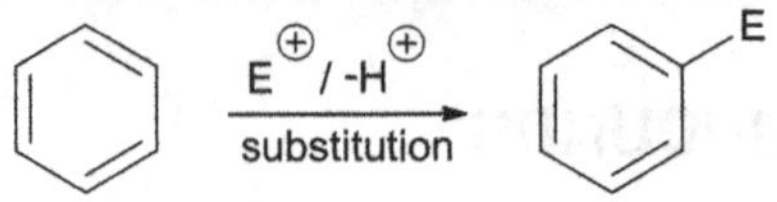

Reaction Type	Conditions	Electrophile
Halogenation	$Br_2/FeBr_3$ or $Cl_2/FeCl_3$	$\overset{\oplus}{Br}$ or $\overset{\oplus}{Cl}$
Nitration	Conc. HNO_3/H_2SO_4	$\overset{\oplus}{N}O_2$
Sulfonation	Conc. H_2SO_4	$\overset{\oplus}{H}SO_3$
Friedel–Crafts alkylation	$RCl/AlCl_3$	$\overset{\oplus}{R}$
Friedel–Crafts acylation	$RCOCl/AlCl_3$	$\overset{\oplus}{R}CO$

7.10.2   Aromatic Diazotization

In Section 7.9.3, you saw how primary amines react with nitrous acid to give diazonium salts. Although alkyl examples are mostly too reactive to be useful, aromatic amines give diazonium salts which are stable at temperatures below 5 °C. These can be used in many different useful reactions. Because aromatic amines can be easily prepared from nitrobenzenes, this gives the practical synthesis shown in Figure 7.49.

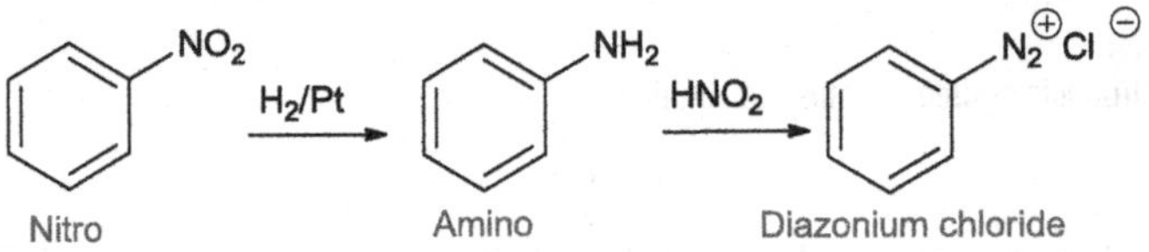

FIGURE 7.49
Formation of aromatic diazonium salts.

The main reaction of diazonium salts is by nucleophilic substitution. This occurs with the loss of a nitrogen molecule. Table 7.6 shows the preparation of substituted aromatics which are not available by direct electrophilic substitution.

Table 7.6	Nucleophilic Substitution of Diazonium Salts	

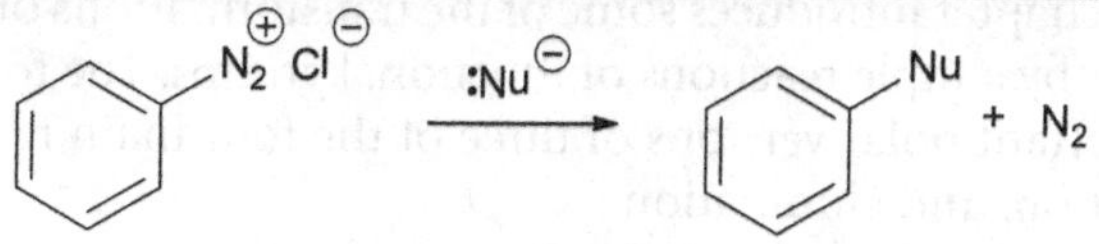

Substituent	Conditions	Product
Cl/Br	CuCl/CuBr	Aryl halide
I	KI	Aryl iodide
CN	CuCN	Aryl nitrile
OH	$H_3O^{\oplus}$	Phenol
H	H_3PO_2	Aryl–H

Diazonium salts are also electrophilic at the terminal nitrogen atom. These salts can react in electrophilic substitutions with other activated aromatic substrates. Usually phenols and amines are used, as shown in Figure 7.50. This reaction is called diazo-coupling and gives highly colored diazo products which are useful as dyes.

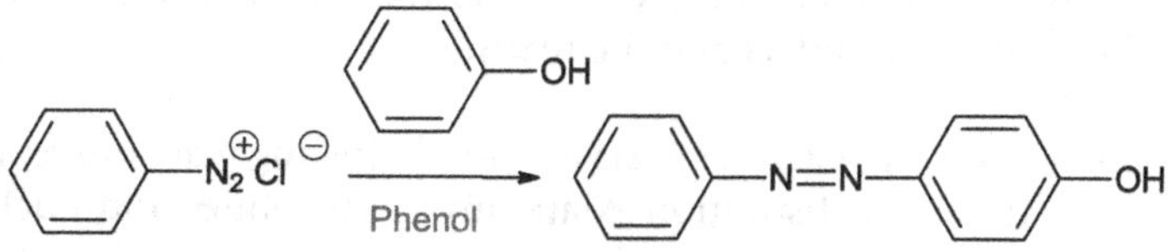

Diazonium chloride   p-Hydroxyazobenzene

FIGURE 7.50
Typical diazo-coupling reaction.

QUESTIONS AND PROGRAMS

This chapter introduces some of the transformations of organic molecules which occur by simple reactions of functional groups. The following Programs explain important polar versions of three of the four main reaction types: substitution, addition, and elimination.

The purpose of these Programs is to show only the main concepts. These concepts can be extended to many different examples. Do not forget that the reaction of one functional group is also the preparation of another. In this way you can make a logical connection between the classes of organic compounds.

The content of this chapter is simply the application of the concepts from the earlier chapters. It is important to know this content before going on. Here are some reminders about the common polar reaction types. These reminders will help you to find the earlier concept topics.

- Reactions have a substrate, reagent(s), and product(s). If you compare the product with the substrate, you will easily identify the type of reaction which has occurred and the reagent needed.
- If you can recognize electron-rich (nucleophilic) and electron-poor (electrophilic) parts of a molecule, you can identify the reacting pairs.
- Reactions occur with bond breaking and making. If you know the factors (inductive effect, resonance) which control the direction of heterolytic cleavage, you will understand the reaction.

Q 7.1   With reference to Chapter 5, classify the following reactions as addition, elimination, or substitution. Mark any electrophiles and nucleophiles in the processes.

(a) $CH_3CH{=}CH_2$ $\xrightarrow{\text{Br}_2}$ $CH_3CH(Br)CH_2Br$

(b) CH_3CHO $\xrightarrow{\text{HCN}}$ $CH_3CH(CN)OH$

(c) $CH_3CH_2CH(Br)CH_3$ $\xrightarrow{\overset{\ominus}{}OH}$ $CH_3CH_2CH{=}CH_2$

(d) $CH_3CH_2CH_2Br$ $\xrightarrow{CH_3C{\equiv}C^{\ominus}}$ $CH_3CH_2CH_2C{\equiv}CCH_3$

(e) $CH_3C{\equiv}CH$ $\xrightarrow{\text{HBr}}$ $CH_3C(Br){=}CH_2$

(f) $CH_3CH_2CO_2CH_3$ $\xrightarrow{CH_3NH_2}$ $CH_3CH_2CONHCH_3$

PROGRAM 19 Nucleophilic Substitution

A In the Programs which follow, curly arrows are used to show bond breaking and making. These arrows will help you follow electron movements and keep a count of electrons. The arrows also show details of the reaction mechanism.

Study the conversion of an alkyl halide into an alcohol.

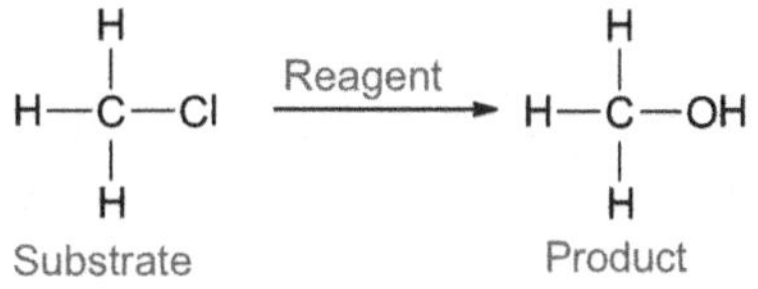

Substrate   Product

If you compare the product with the substrate, you see that the chlorine ligand is replaced by a hydroxyl ligand. The process must be a substitution. However, no information is given about how the reaction occurred.

✎ Use your knowledge of bond polarity (inductive effect) to draw a reasonable mechanism for the substitution process.

B The C–Cl bond is the most polar bond in the substrate. This means the electrophilic carbon can be attacked by a nucleophile. By studying the alcohol product, we can see that hydroxide is the likely reagent (see Program 15**E**).

$$HO^{\ominus} \quad CH_3{-}Cl \longrightarrow CH_3{-}OH \; + \; Cl^{\ominus}$$

You have just described a nucleophilic substitution. The reaction is classified as nucleophilic substitution because a nucleophile replaces the chloride leaving group. Proper choice of the nucleophile can give many different functional classes of product.

All of these substitutions are the simple exchange of single bonded ligands, and there is no change in the bond order between substrate and product.

✎ Can you suggest some examples of substitution reactions?

C You may have included nucleophiles such as:

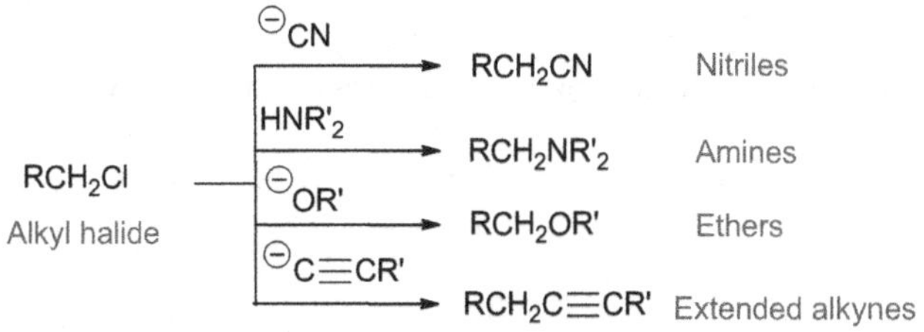

$^{\ominus}CN$	RCH_2CN	Nitriles
HNR'_2	$RCH_2NR'_2$	Amines
$^{\ominus}OR'$	RCH_2OR'	Ethers
$^{\ominus}C{\equiv}CR'$	$RCH_2C{\equiv}CR'$	Extended alkynes

RCH₂Cl
Alkyl halide

There are two ways in which the above reactions might occur. The difference depends on the timing of the departure of the leaving group and the arrival of the nucleophile.

Continued...

–Cont'd

Part **B** showed one of these options in which the two processes occur at the same time. This is classified as a one-step S_N2 process. This process depends on the concentration of both the nucleophile and the substrate in the rate-determining step.

✎ Can you draw a second possible description for the process?

D   Another description is if the leaving group departs first, without help from the nucleophile. This heterolysis gives a carbocation. Then, the nucleophile reacts with this carbocation to complete the substitution.

$$R_3C—Br \xrightarrow{\text{-Br}^{\ominus}} R_3C^{\oplus} \quad H_2O \overset{\text{-H}^{\oplus}}{\longrightarrow} R_3C—OH$$

This second description is classified as a two-step S_N1 reaction. This mechanism depends only on the substrate concentration in the rate-determining step of heterolytic ionization.

Although we are not limited to alkyl halides, they are very good substrates for nucleophilic substitution. The main reason is that halides are good leaving groups. For instance, study the reverse reaction of the one that is described in **B**.

$$Cl^{\ominus} \quad CH_3—OH \longrightarrow CH_3—Cl \ + \ ^{\ominus}OH$$

✎ Is this a likely reaction?

E   As shown, the substitution will not work. Because chloride is the weak conjugate base of a strong acid, HCl, it will not displace hydroxide, which is the strong conjugate base of a weak acid, H_2O.

To produce the alkyl halide product, it is necessary to change the hydroxide into a better leaving group. This is done by acid catalysis. The hydroxyl group is protonated and can leave as a neutral water molecule. Therefore, the reaction is done in concentrated acid medium.

$$Cl^{\ominus} \quad CH_3—\overset{\oplus}{O}H_2 \longrightarrow CH_3—Cl \ + \ H_2O$$

✎ Can you draw a pathway of the steps in the above conversion?

F The first step of the mechanism is the protonation of the alcohol hydroxyl group. Depending on the particular alcohol R-group, either an S_N1 or S_N2 type of conversion follows.

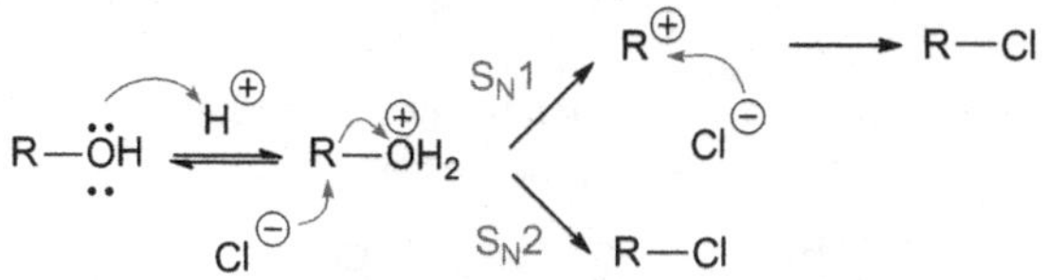

The use of an acid catalyst is often used to improve the leaving group of certain strong conjugate bases (nucleophiles).

Q 7.2 Complete the following nucleophilic substitution reaction equations by giving either the product or an appropriate reagent. In each case, give the functional class of both substrate and product.

(a) 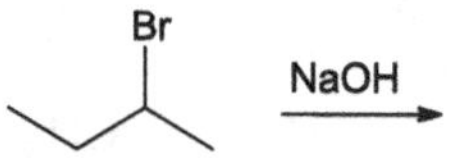   NaOH

(b) 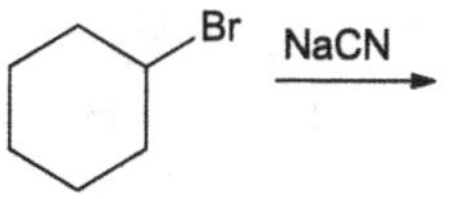   NaCN

(c)

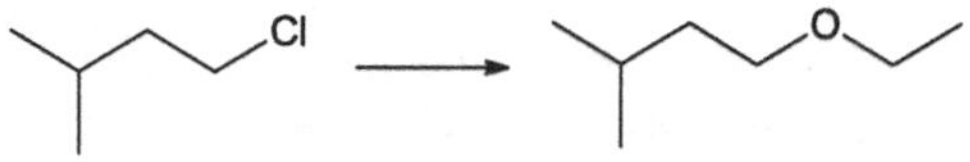

(d)   CH₃I   (CH₃)₃N

(e)

Q 7.3 The S_N2 reaction of 1-bromobutane with NaOH occurs readily. What happens to the rate of this reaction if
(a) the concentration of NaOH is doubled.
(b) the concentrations of both NaOH and 1-bromobutane are doubled.
(c) the volume of the reaction solution is doubled.

PROGRAM 20 Elimination

A Program 19 showed nucleophilic substitution at a saturated carbon atom. The nucleophile attacked the saturated center with an electron pair and displaced a suitable leaving group.

There is another reaction which can occur if the nucleophilic electron pair acts as a base. Study again the reaction of an alkyl halide with hydroxide.

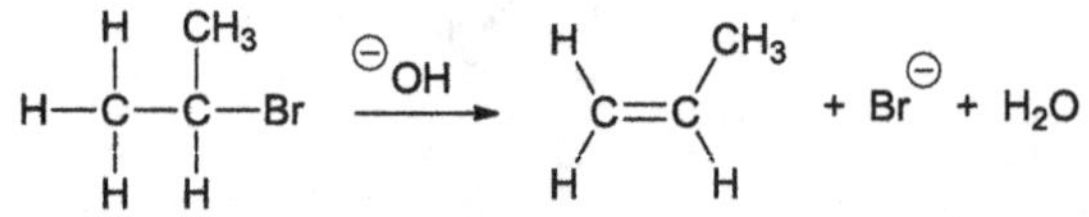

The overall reaction is elimination. It is a 1,2-elimination between adjacent centers. This loss of a proton and a bromide causes an increase in the bond order between the original carbon centers.

✎ Can you rewrite the above reaction with a curly arrow description?

B The detailed equation should look like the following:

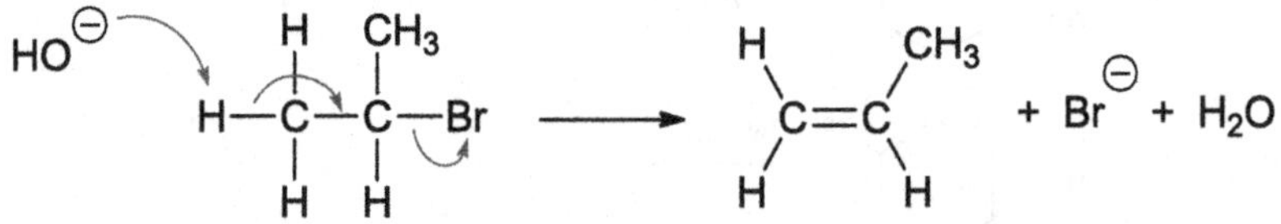

Like nucleophilic substitution, the elimination reaction can follow two different pathways. This is again controlled by the exact order of events. If the leaving group leaves by ionization to give a carbocation, the mechanism is called E1. If the leaving group is displaced in a single step when the base removes an adjacent proton, the mechanism is called E2. This is the mechanism shown above.

Like substitution reactions, E1 shows that the rate-determining step is controlled only by the concentration of substrate. In E2, the rate-determining step is controlled by both the concentration of substrate and base.

✎ Use the above example to draw the alternative E1 sequence.

C You should have drawn the following.

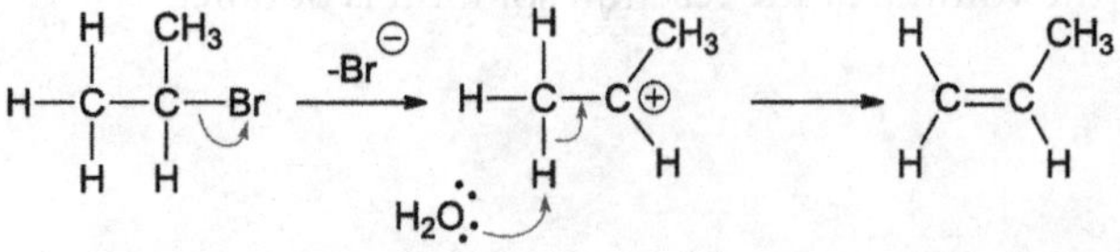

Q 7.4 Draw structures for the alkene(s) which are produced by E2 elimination
if the following alkyl halides are treated with sodium ethoxide (NaOEt).

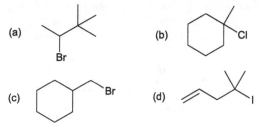

(a)

(b)

(c)

(d)

PROGRAM 21 Nucleophilic Addition

A Substitution does not change the bond order, but elimination causes an increase
in bond order. Both of these reaction types have a saturated center which has a leaving
group, and both proceed by reaction with an electron pair donor nucleophile/base.

An addition reaction is the opposite of elimination. Therefore it must cause a decrease in
the bond order. Addition can follow two main mechanisms. Which mechanism is followed
depends on the type of unsaturation (double bond). The double bond may be the dipolar
type found in carbonyl groups, or largely non-polar as found in alkenes and alkynes.

A dipolar carbonyl group has both electrophilic and nucleophilic centers. Nucleophilic
attack needs an electrophilic center. The example reaction of cyanide with an aldehyde
shows the general outcome (see Program 15**C** and **D**).

Cyanohydrin

Because the reaction is started by a nucleophile, it is called nucleophilic addition. The
main feature of addition is the decrease in bond order. This happens when two new
single bonds form to complete the valency requirements.

There are many different nucleophiles, and many different addition products can be
made. Nucleophilic additions can occur under acid or base conditions. This depends
on the exact nucleophile–electrophile pair in the reaction.

For example, the addition reaction of water to aldehydes or ketones to give hydrates
can be done under basic conditions in a way similar to the cyanohydrin formation.

Hydrate

✎ Can you draw the acid-catalyzed process?

B   You should have drawn something like the following. Initial protonation (acid–base reaction) of the carbonyl oxygen gives a resonance stabilized oxonium ion. The positive charge on oxygen makes the carbonyl carbon more electrophilic (see the second resonance contributor). Then it is attacked by the weak nonionic nucleophile water.

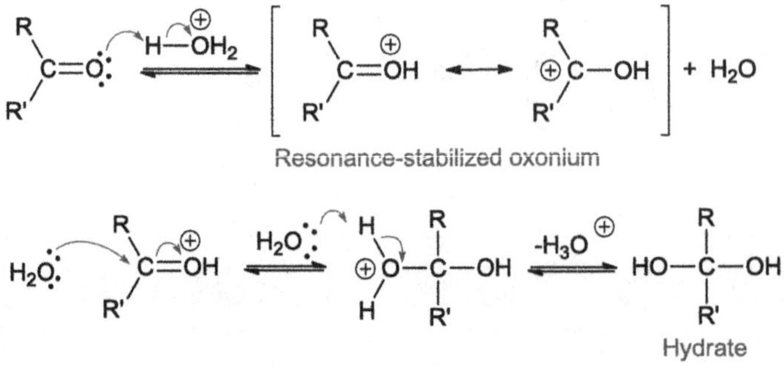

Resonance-stabilized oxonium

Hydrate

This type of nucleophilic attack at the polarized carbonyl group can be done with various nucleophiles to give many different useful products.

The product of addition reaction can often react further under the same conditions. An important example is the above hydration reaction, using an alcohol ROH instead of water.

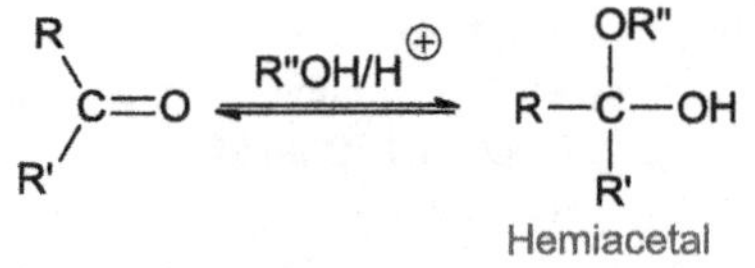

Hemiacetal

This reaction follows the same pathway as for the addition of water.

✎ Can you draw this sequence with curly arrows?

C   You should have drawn the following:

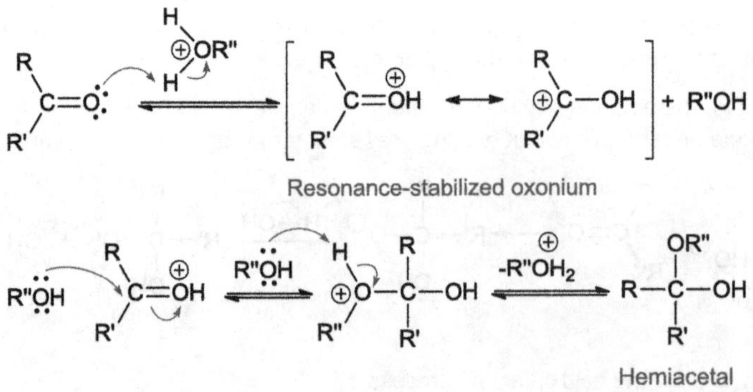

Resonance-stabilized oxonium

Hemiacetal

─Cont'd

> Because the reaction mixture is still acidic, equilibrium protonation of the basic oxygens in the hemiacetal can occur. Protonation of the hydroxyl group gives a good leaving group which is easy to replace with a second alcohol molecule. This is similar to the acid-catalyzed substitution in Program 19**E** and **F**.
>
> ✎ Try to draw a mechanism for this nucleophilic substitution step.

D   Following Program 19, you should have drawn:

Acetal   Resonance-stabilized
                                          alkyloxonium

Because they are acid-catalyzed, the reaction steps are drawn as equilibria. The reactions can be reversed by changing the concentration of the reaction components.

E   Reactions with primary amino nucleophiles RNH$_2$ are related to the addition reactions seen in **D**. Initial nucleophilic addition occurs in exactly the same way as it does with alcohols. It can be written as:

Resonance-stabilized oxonium ion

The difference between this and the earlier hemiacetal intermediate is that the trivalent nitrogen still has a further hydrogen attached. Therefore, the molecule has the substituents which are needed for an elimination step to occur.

✎ Try to draw a representation of this step.

F   You should have identified the elimination of a water molecule in the following way.

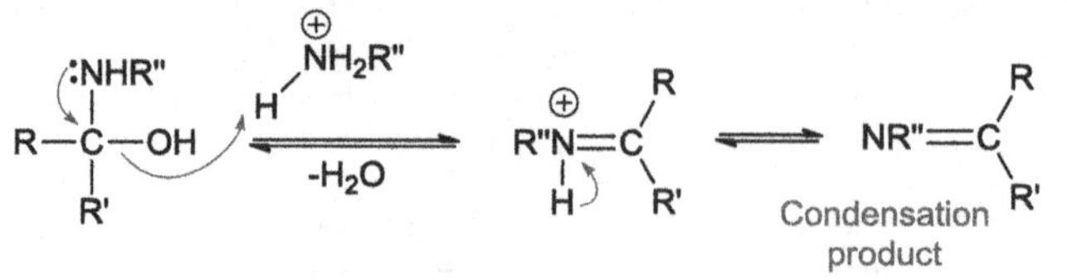

Although the steps in **E** and **F** can be written in slightly different ways, the overall process shows a nucleophilic addition which is followed by an elimination.

This combination is called a condensation reaction. It leads to a number of useful products with primary amino nucleophiles which have different substituents. The overall reaction is the substitution of a C=O with a C=NR.

Q 7.5   Complete the following nucleophilic addition reaction equations by drawing the missing carbonyl substrate, reagent, or product. In each case, show the functional class of the product.

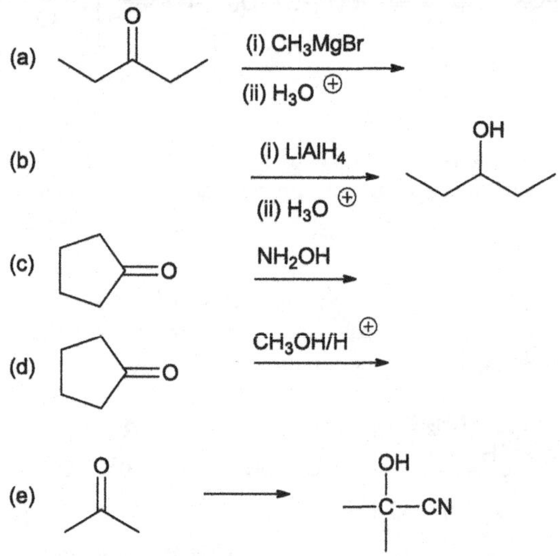

(a)   (i) CH$_3$MgBr
                              (ii) H$_3$O $^{\oplus}$

(b)   (i) LiAlH$_4$
      (ii) H$_3$O $^{\oplus}$

(c)   NH$_2$OH

(d)   CH$_3$OH/H $^{\oplus}$

(e)   ⟶

PROGRAM 22 Nucleophilic Acyl Substitution

A   Not only aldehydes and ketones have a carbonyl group. There are many other acyl derivatives. All of them have the common feature of the carbonyl group in combination with another functionality.

Study the nucleophilic attack on acyl halides (alkyl halide combined with a carbonyl group).

Addition   Elimination

Tetrahedral   Acyl substitution
intermediate   product

The first step of nucleophilic addition goes as we expect. It gives the tetrahedral intermediate. However, instead of adding a proton to complete the addition, the good halide leaving group is eliminated.

The overall reaction is a substitution. The result seems the same as the S_N reactions which we studied in Program 19, but the mechanism is different. It is a combination of the steps of addition and elimination.

Aldehydes and ketones do not react in acyl substitution. This is because they do not have a good leaving group for the elimination step. Instead, they have only very poor leaving group options of hydride or a carbanion.

✎ Try to draw the sequence for the reaction of an ester with ammonia.

B   The ester is the acyl derivative, and ammonia is the nucleophile. The substitution reaction gives an amide. You should have drawn something like the following:

Ester   1°Amide

These reactions move forward because of the stability of the product relative to the substrate. The direction of reaction direction is controlled by the strength of the nucleophile and the quality of the leaving group. This gives the general order of acyl derivative reactivity as: acyl halide > acid anhydride > ester > acid > amide.

Q 7.6   Draw structural formulae for the following acyl derivatives.
  (a) Methyl 3-methylbutanoate
  (b) Octanonitrile
  (c) Phenylacetic anhydride
  (d) N-Cyclohexylethamide

Q 7.7   Give structural formulae and the compound class name for the products from treatment of butanoyl chloride, $CH_3CH_2CH_2COCl$, with the following reagents.
  (a) Water
  (b) N,N-Dimethylamine
  (c) Lithium ethanoate
  (d) Cyclohexanol

PROGRAM 23 Electrophilic Addition

A   Program 21 covered nucleophilic addition reactions. Nucleophilic attack was caused by the dipolar nature of the unsaturated bond, usually a carbonyl group.

The all-carbon unsaturated bond equivalents are alkenes and alkynes. They are non-polar by comparison. The multiple bond sites are in fact electron rich areas. Because of this, they can act as nucleophiles. These nucleophiles react with electrophilic reagents. Study the following reaction.

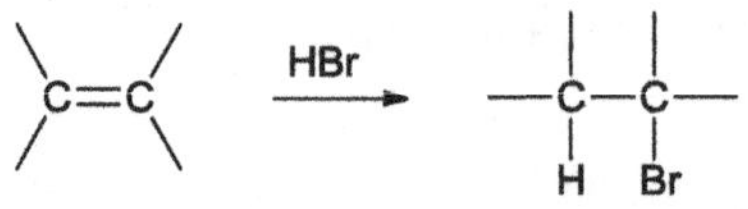

The reaction is clearly addition, and the bond order decreases as expected. The order of events is the opposite of the order of addition to carbonyl groups. Therefore it is classified as electrophilic addition. See Section 7.3.1 and Program 15**A**–**C**.

✎ Draw a reasonable mechanism for the above reaction.

B   You should have drawn the following.

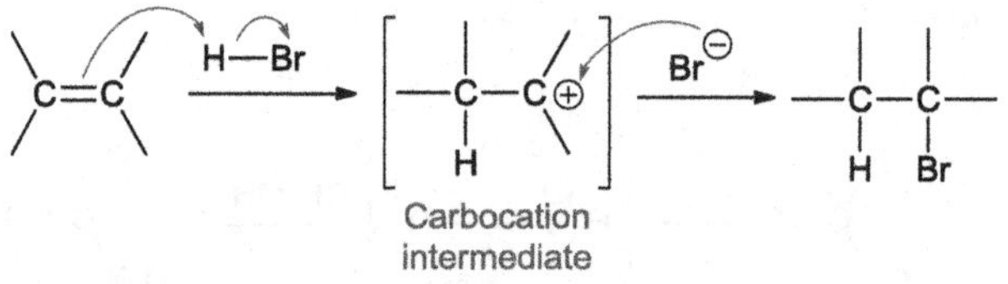

Carbocation
intermediate

This shows the initial reaction of the nucleophilic multiple bonds with the electrophilic end of the addition reagent. In this case, it is a proton. This reaction gives an intermediate carbocation. The reaction is completed by bonding the nucleophilic end of the reagent. In this case a bromide.

In the example above, the product is an alkyl halide. This class of compound was shown as a flexible substrate for nucleophilic substitution (Program 19) and elimination reactions (Program 20). Addition is the reverse of elimination. By using suitable pairs of electrophilic and nucleophilic addition reagents, many different products can be made from alkenes and alkynes.

Study the following sequence. This is a special example which does not follow the electrophile-initiated addition at an unsaturated carbon center.

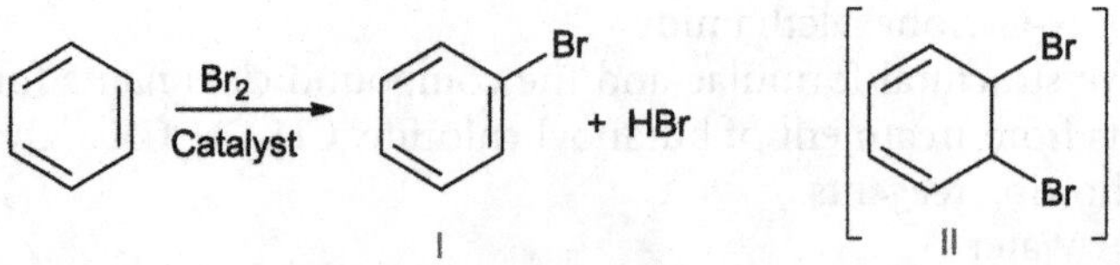

✎ Explain why a C=C would prefer to react by substitution to give **I** instead of addition to give **II** as you might expect.

C Remember that an aromatic system gets large resonance stabilization from its conjugated system. This stabilization is lost if an addition reaction occurs. This happens because the double bond conjugated system would be broken. Therefore, after the first step of electrophilic addition, the intermediate carbocation eliminates a proton rather than adding the nucleophilic half of the addition reagent. This reforms the conjugated system. See Section 7.10 and Program 24.

PROGRAM 24 Electrophilic Aromatic Substitution

A Many different substituents may be introduced onto the **aromatic** ring by direct electrophilic substitution. What is the mechanism for this general reaction? Study the bromination of benzene. Because the benzene nucleus is electron rich, it can act as the nucleophile.

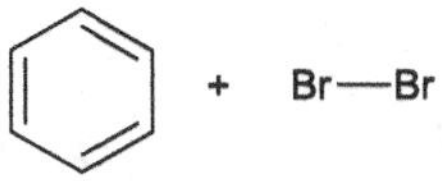

✎ Draw diagrams to show this initial nucleophilic attack.

B

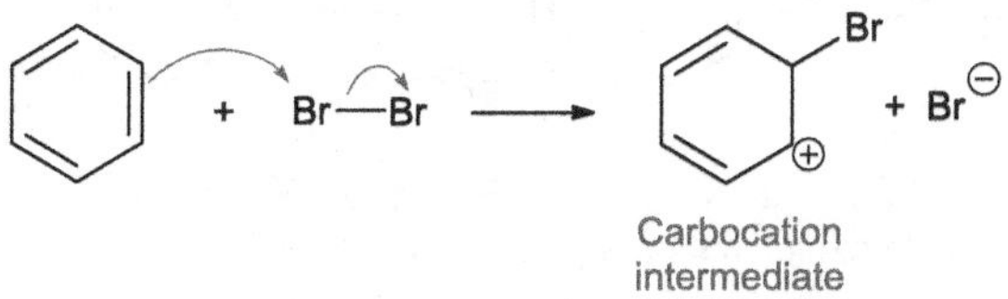

Carbocation intermediate

This carbocation is stabilized by delocalization.

✎ Draw the resonance forms and hybrid for this carbocation.

C

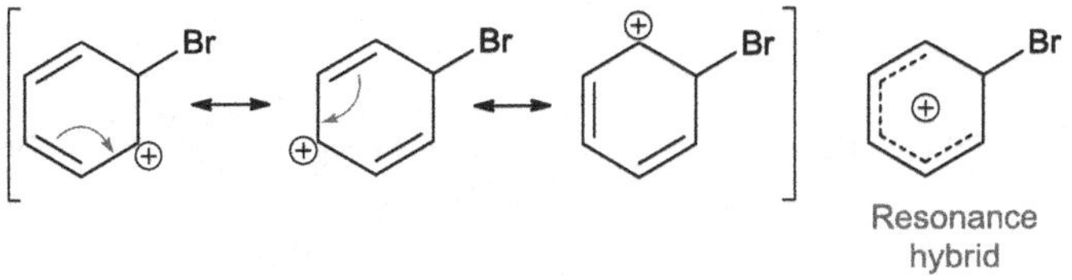

Resonance hybrid

If the bromide ion simply undergoes an addition reaction as with simple alkenes, a non-aromatic product would result.

✎ Draw this possible step.

D

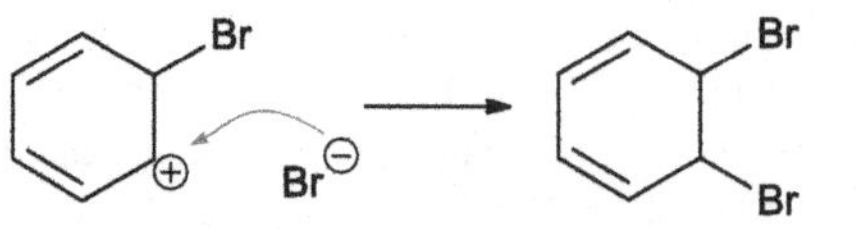

Non-aromatic
addition product

This gives a non-aromatic product of higher energy. It is better if the bromide ion acts as a base in the deprotonation reaction to regenerate the delocalized aromatic system.

✎ Try to draw this step.

E

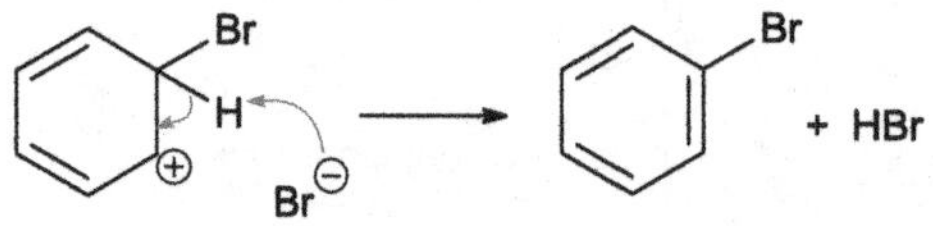

+ HBr

Substitution product

Therefore the full sequence may be combined as follows.

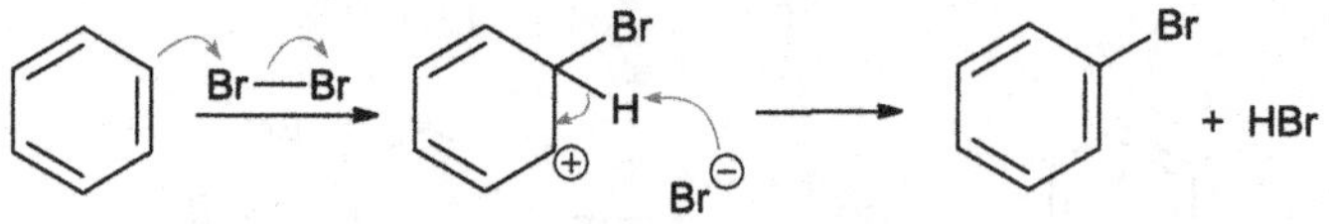

+ HBr

If you refer to Program 22, you will see that this reaction is an electrophilic version of the nucleophilic addition/elimination sequence which occurs with acyl derivatives.

Usually, because of the relatively low nucleophilic character of the aromatic nucleus, the quality of the electrophile is often increased by a Lewis acid catalyst. For the above bromination, this is done with $FeBr_3$.

✎ How might this Lewis acid activate the Br_2?

F

We can expect similar activation for chlorination ($Cl_2/FeCl_3$), alkylation ($RHal/AlHal_3$), and acylation ($RCOHal/AlHal_3$).

✎ Draw the formation of these reactive species. Use examples of your choice for alkylation and acylation.

G

$:\overset{..}{\underset{..}{Cl}}-\overset{..}{\underset{..}{Cl}}: + FeCl_3 \longrightarrow :\overset{..}{\underset{..}{Cl}}\text{------}\overset{..}{\underset{..}{Cl}} :FeCl_3 \longrightarrow \overset{\oplus}{Cl} + \overset{\ominus}{FeCl_4}$

$\underset{H_3C}{\overset{Br}{\underset{}{\overset{|}{C}H}}}\overset{}{\underset{CH_3}{}} \xrightarrow{AlBr_3} \underset{H_3C}{\overset{H}{\underset{CH_3}{\overset{|}{\underset{\oplus}{C}}}}} + \overset{\ominus}{AlBr_4}$

$\underset{H_3C}{\overset{O}{\overset{\|}{C}}}\overset{}{\underset{Cl}{}} \xrightarrow{AlCl_3} \left[CH_3\overset{\oplus}{C}=O \longleftrightarrow CH_3C\overset{\oplus}{=}O \right] + \overset{\ominus}{AlCl_4}$

Other common substitution electrophiles for sulfonation and nitration are prepared as follows:

$H_2SO_4 + SO_3 \rightleftharpoons \overset{\oplus}{SO_3H} + \overset{\ominus}{HSO_4}$

$HNO_3 + H_2SO_4 \rightleftharpoons \overset{\oplus}{NO_2} + \overset{\ominus}{HSO_4} + H_2O$

Q 7.8 Draw a structure for all of the possible carbocations from the addition of a proton to the following alkenes. Label each as 1°, 2°, or 3° and mark the most stable carbocation for each alkene.

(a) $\underset{CH_3CH_2}{\overset{CH_3}{C}}=CHCH_3$

(b) $CH_3CH=CHCH_2CH_3$

(c) (cyclopentene)—CH_3

(d) (cyclohexane ring)=CH_2

Q 7.9 Predict the major products from the reaction of 2-methyl-1-pentene with the following reagents.
(a) H_2O/H_2SO_4
(b) Br_2
(c) HBr

Q 7.10 Draw structures for the products formed by the treatment of propanal with the following reagents.
(a) $LiAlH_4$ followed by water
(b) H_2/Pt catalyst
(c) Acidified $K_2Cr_2O_7$
(d) $HOCH_2CH_2OH$/catalytic H_2SO_4

(e) $CH_3CH_2NH_2$/acid catalyst
(f) $(NH_3)_2Ag^+$

Q 7.11 Supply reagents/conditions for the reactions shown below.

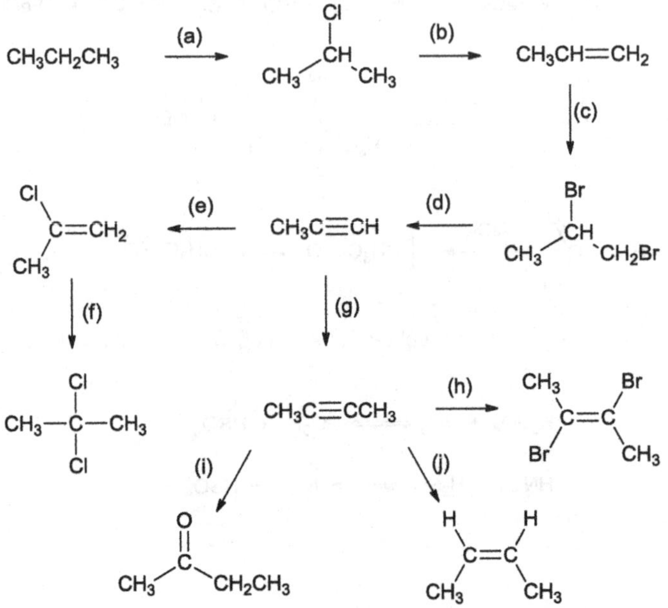

Q 7.12 Supply structures and reaction types for each of the following one-step reactions. In each case name all the compound functional classes.

(a) $(CH_3)_2CHCH_2CHO$ $\xrightarrow{\text{LiAlH}_4}$

(b) $(CH_3)_2CHCH_2Br$ $\xrightarrow[\text{dilute}]{\overset{\oplus}{Na}\ \overset{\ominus}{O}CH_2CH_3}$

(c) CH_3CH_2OH $\xrightarrow{\text{PCl}_3}$ $\xrightarrow{\text{CH}_3\text{OH}}$

(d) CH_3CH_2CHO $\xrightarrow[\text{H}_2\text{SO}_4]{\text{NaCN}}$ $\xrightarrow{\text{heat}}$

Q 7.13 Describe simple chemical tests which will give a visual way to distinguish between the following pairs of compounds. In each case record what is seen, and write an equation for each positive test.
(a) $CH_3CH=C(CH_3)_2$ and $CH_3CH_2CH(CH_3)_2$
(b) $CH_3CH_2CH(OH)CH_2CH_3$ and $CH_3CH_2CH_2CH(OH)CH_3$
(c) $(CH_3)_2CHCO_2H$ and $(CH_3)_2CHCHO$
(d) $CH_3CH_2CH_2NH_2$ and $CH_3CH_2CONH_2$
(e) Cyclohexanol and cyclohexanone
(f) Bromocyclopentane and cyclopentanol

Q 7.14 Study the list of reactants, reagents, and products and choose the letter which is the best one to complete the reactions below.

[A] $CH_3CH_2CH_2CO_2H$ [B] HCl

[C] $(CH_3)_2C(OH)C(OH)(CH_3)_2$ [D] $(CH_3)_2C=O$

[E] NaCl   [F] $CH_3CH_2CH(OH)CO_2H$

[G] $(CH_3)_2CHCOCl$ [H] CH_3CH_2CHO

[I] (i) $LiAlH_4$ (ii) H_3O^+ [J] PCl_3

[K] H_2SO_4/heat   [L] (i) I_2/NaOH (ii) H_3O^+

[M] $(CH_3)_2CHCO_2H$ [N] $NaNO_2$/HCl

(a)

$$\underset{CH_3}{\overset{CH_3}{}}C=C\underset{CH_3}{\overset{CH_3}{}} \quad \xrightarrow[\text{(ii) Zn/H}_2\text{O}]{\text{(i) O}_3}$$

(b) $CH_3CH_2CO_2H \longrightarrow CH_3CH_2COCl$

(c)

$$\underset{CH_3}{\overset{CH_3CH_2}{}}C=O \longrightarrow CH_3CH_2CO_2H + CHI_3$$

(d) $CH_3CH_2CO_2H \longrightarrow CH_3CH_2CH_2OH$

(e) $CH_3CH_2CHO \xrightarrow[\text{(ii) H}_3\text{O}^{\oplus}]{\text{(i) HCN}}$

(f) $\xrightarrow{CH_3NH_2}$ $CH_3\underset{CH_3}{\overset{H}{\underset{|}{\overset{|}{C}}}}\overset{\overset{O}{\|}}{C}-NHCH_3$

Q 7.15 Study the list of reactants, reagents, and products and choose the letter(s) which is/are the best to complete the reactions below.

[A] $CH_3CH_2CH_2CHO$ [B] $CH_3CH=C=CH_2$

[C] $CH_3CH(I)CH_3$ [D] $CH_3CO_2^-$

[E] $CH_3CH_2CO_2H$ [F] $CH_3CH_2CH(Br)CH_3$

[G] CH_3CO_2H [H] $CH_3CH_2CH_2CH_2Br$

[I] CH_3CH_2OH/H^+ [J] CHI_3

[K] $CH_3CH_2COCH_3$ [L] $CH_3CH_2C\equiv CH$

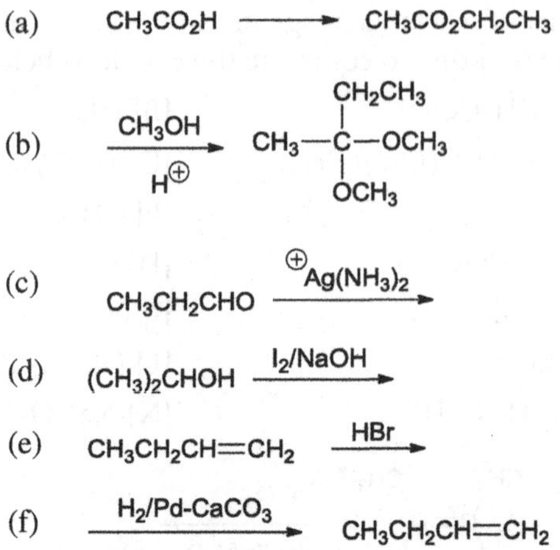

(a) $CH_3CO_2H \longrightarrow CH_3CO_2CH_2CH_3$

(b) $\xrightarrow[H^{\oplus}]{CH_3OH}$ $CH_3-\underset{\underset{OCH_3}{|}}{\overset{\overset{CH_2CH_3}{|}}{C}}-OCH_3$

(c) $CH_3CH_2CHO \xrightarrow{\overset{\oplus}{Ag(NH_3)_2}}$

(d) $(CH_3)_2CHOH \xrightarrow{I_2/NaOH}$

(e) $CH_3CH_2CH{=}CH_2 \xrightarrow{HBr}$

(f) $\xrightarrow{H_2/Pd\text{-}CaCO_3} CH_3CH_2CH{=}CH_2$

Q 7.16 Study the list of reagents/reaction conditions and choose the best one or combination which completes the reaction below.

[A] $NaNH_2$	[B] (i) O_3 (ii) Zn/HCl	[C] CH_3Br
[D] PCl_5	[E] $NH_3/Heat$	[F] H_3O^+
[G] $c.H_2SO_4/140°$	[H] $c.H_2SO_4/170°$	[I] Excess $I_2/NaOH$
[J] $LiAlH_4$	[K] HCN	[L] $Pd\text{-}CaCO_3/H_2$
[M] $d.H_2SO_4/HgSO_4$	[N] $Pd\text{-}C/H_2$	[O] $Na_2Cr_2O_7/H^+$

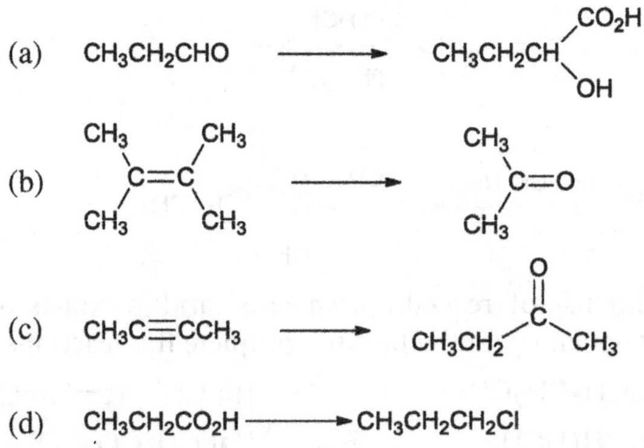

(a) $CH_3CH_2CHO \longrightarrow CH_3CH_2CH\overset{\overset{CO_2H}{\diagup}}{\underset{\underset{OH}{\diagdown}}{}}$

(b) $\underset{\underset{CH_3}{}}{\overset{\overset{CH_3}{}}{C}}{=}\underset{\underset{CH_3}{}}{\overset{\overset{CH_3}{}}{C}} \longrightarrow \underset{\underset{CH_3}{}}{\overset{\overset{CH_3}{}}{C}}{=}O$

(c) $CH_3C{\equiv}CCH_3 \longrightarrow CH_3CH_2\overset{\overset{O}{\parallel}}{C}CH_3$

(d) $CH_3CH_2CO_2H \longrightarrow CH_3CH_2CH_2Cl$

Q 7.17 Use 1-butyne as starting material and show how the following products may be made.

(a) $CH_3CH_2C{\equiv}CCH_3$ (b) $CH_3CH_2CO.CH_3$ (c) CH_3CH_2CHO

Q 7.18 Study the following reaction:

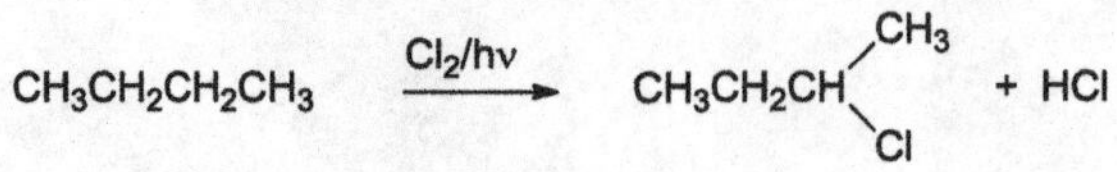

$CH_3CH_2CH_2CH_3 \xrightarrow{Cl_2/h\nu} CH_3CH_2CH\overset{\overset{CH_3}{\diagup}}{\underset{\underset{Cl}{\diagdown}}{}} + HCl$

    (a) Classify the reaction type.

    (b) Is the product chiral? Why?

    (c) Write a mechanism for the reaction as shown.

Q 7.19 Explain why the following reaction sequence is not favorable:

$$CH_4 + Cl_2 \xrightarrow{\text{hv}} CH_3Cl + HCl$$

Give a better solution.

Q 7.20 Show how the following interconversions can be performed. More than one step may be required.

    (a) $CH_3CH_2CH_2CHO \rightarrow CH_3CH_2CH_2CH_2OCH_3$

    (b) $CH_3CH_2CO_2CH_3 \rightarrow CH_3CH_2CH_2OH$

    (c) $CH_3CH_2OCH_3 \rightarrow CH_2{=}CH_2$

    (d) $CH_3CH_2CH_2CO \cdot CH_3 \rightarrow CH_3CH_2CH_2CO_2H$

Q 7.21 The chiral substance **A** (C_6H_{12}) reacts with HBr to give a chiral second-ary alkyl halide **B** $(C_6H_{13}Br)$. Alternatively, if A is treated with Pt/H_2 (hydrogenation), it gives **C** (C_6H_{14}) which is not chiral. Write a reaction scheme to show the structures of **A–C**.

Q 7.22 A symmetrical ether $C_6H_{14}O$ is treated with excess boiling HI to give a secondary alkyl halide C_3H_7I. This iodide reacts with NaCN to give a product which, on warming with dilute acid, gives 2-methylpropanoic acid. Draw the reactions which have occurred.

Q 7.23 Suggest a method for the following multistep transformation:

$$CH_3CH_2CH(CH_3)CH_2OCH_2CH(CH_3)CH_2CH_3 \rightarrow CH_3CH_2CH(OH)CH_3$$

Q 7.24 The compound **A** (C_3H_6O) reacts with aqueous NaCN and H_2SO_4 to give a racemic cyanohydrin **B**. Hydrolysis of B with aqueous acid affords **C**, and further, dehydration of C yields **D** which decolourizes a Br_2 test solution. D also reacts with MeOH (excess)/H_2SO_4 to give **E** $(C_5H_8O_2)$. Identify the compounds **A–E**.

Q 7.25 An aldehyde **A** $(C_5H_{10}O)$ can be reduced by $LiAlH_4$ to **B**. Treatment of B with concentrated H_2SO_4 at 180 °C gives an alkene **C**. Ozonolysis of C gives methanal and **D** (C_4H_8O), which has a positive iodoform reac-tion. Identify **A–D**.

Q 7.26 The compound **A** (C_5H_{10}) undergoes ozonolysis to **B** and methanal. **B** when treated with $NaOH/I_2$ followed by aqueous acid gives a yellow precipitate and **C** $(C_3H_6O_2)$. C bubbles when it is treated with $NaHCO_3$. Identify **A–C**.

Q 7.27 A chiral amide **A** $(C_6H_{11}NO)$ decolorizes a Br_2 test solution. A can be hydrolyzed to an acid **B** $(C_5H_8O_2)$ which on ozonolysis gives **C** $(C_4H_6O_3)$ which is still chiral. Identify **A–C** and write a reaction sequence for the conversions.

Q 7.28 The three isomeric dibromobenzenes were each mononitrated. One gave a single product, another gave two products, and the third gave three products. Write equations to explain these results.

CHAPTER 8
Natural Product Biomolecules

8.1 WHAT ARE BIOMOLECULES?

Natural products refer to the many different compounds which are made by living organisms in the world around us. The history of organic chemistry is linked to these naturally occurring molecules. The purpose of this chapter is to look at some of the important types of these natural products. With the concepts from earlier chapters, we show the link between organic chemistry and biochemistry—the composition, structure, and reactions of the biomolecules in living systems.

Biomolecules have the same functional groups that we studied in earlier chapters. However, biomolecules can have more complex structures with many functional groups in the same molecule. The individual functional group properties, structure, and basic principles which apply to simple organic molecules are still mostly the same in these natural systems.

This chapter covers four classes of natural products: carbohydrates (sugars), lipids, peptides/proteins (from amino acids), and nucleic acids. No biochemical details are shown, and we discuss only the properties and principles which were covered in the earlier material. Systematic naming is used only if it is needed to recognize molecular classes.

8.2 CARBOHYDRATES

Carbohydrates are polyhydroxy aldehydes and ketones, or their more complex precursors such as starch or cellulose. They are found in all living organisms. Some simple examples are shown in Figure 8.1. See Program 25.

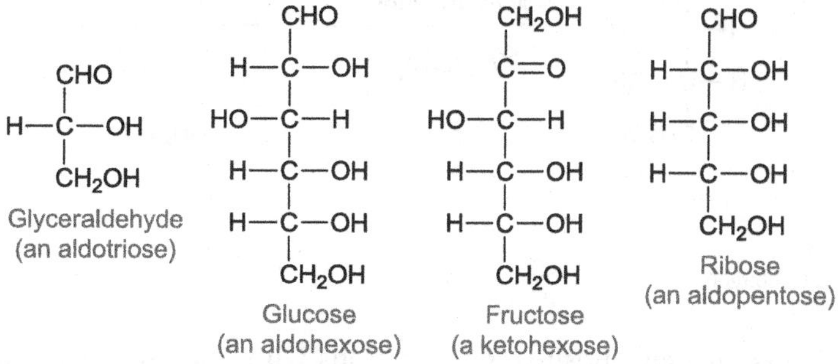

FIGURE 8.1
Diagrams of some simple sugars (*Fischer projections*).

Organic Chemistry Concepts: An EFL Approach. http://dx.doi.org/10.1016/B978-0-12-801699-2.00008-0

Carbohydrates are classed as either simple or complex. They are simple or complex depending on whether or not there are one or more sugar units in the structure. Simple sugars, as in Figure 8.1, are the monosaccharide building blocks for complex carbohydrates—disaccharides, trisaccharides … polysaccharides.

Monosaccharides are subdivided into aldoses or ketoses, based on whether there is an aldehyde or ketone as the carbonyl group. The number of carbons in the sugar backbone is indicated with a prefix such as tri-, tetr-, pent-, and hex-. See the examples in Figure 8.1.

As shown in Figure 8.2, carbohydrates are the products of photosynthesis and are Nature's energy source.

photosynthesis	energy is stored

$$6CO_2 + 6H_2O \xrightarrow[\text{chlorophyll}]{h\nu} C_6H_{12}O_6 + 6O_2$$

glucose

energy is released	metabolism

FIGURE 8.2
Glucose synthesis and metabolism.

The link to simple organic molecules is through the functional groups such as hydroxy from alcohols, and carbonyl groups from aldehydes and ketones. In addition, there can be chirality centers. These functional groups and stereochemical features have the same properties described earlier for their simpler equivalents.

For example, aldoses, like any aldehyde, can be oxidized to the carboxylic acid (aldonic acid). The Tollens' silver mirror test or the copper(II)-based Fehling's and Benedict's reagents can be used to show whether a simple sugar is present. The metal reduction product of elemental silver, as in Figure 8.3, or red copper(I) oxide give clear visual results.

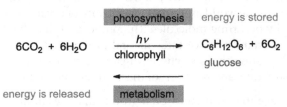

Glucose
(an aldohexose)

Gluconic acid
(an aldonic acid)

FIGURE 8.3
Tollens' aldose oxidation.

These reagents also give positive test reactions with 2-ketoses. This surprising result can be explained by the base-catalyzed tautomerism shown in Figure 8.4, which gives the corresponding aldose. This aldose gives a positive aldehyde test. The equilibrating species differ only in the position of the tautomeric hydrogen.

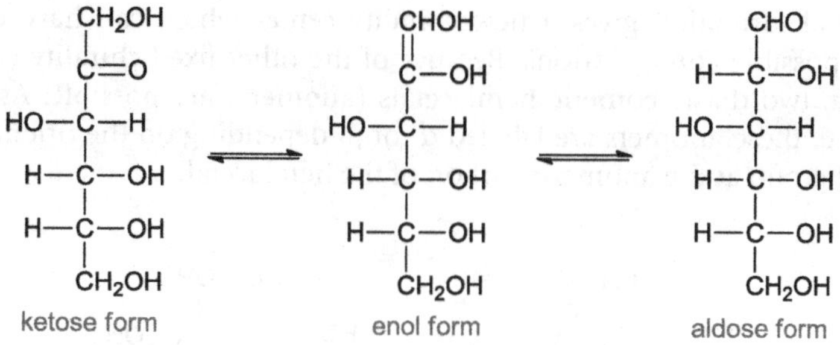

FIGURE 8.4
Ketose to aldose tautomerism.

The oxidation of the carbonyl group to a carboxylic acid causes the reduction of the oxidizing agent. This is the reason that monosaccharides are often called reducing sugars.

The carbonyl groups in sugars can be reduced by the hydride reagents shown in Section 7.7.1.3. The alcohol groups react in the expected ways such as the ether formation, seen in Section 7.6, or esterification, seen in Section 7.8.2.

Sugars can be in open chain or cyclic structures. As Figure 8.5 shows, these structures come from reversible nucleophilic addition of the alcohol and carbonyl groups in the sugar to give hemiacetals. This reaction is an **intramolecular** version of the earlier intermolecular one. The only difference from the earlier examples in Section 7.7.1.4 is that both reacting species are in the same molecule. Six-membered pyranose and five-membered furanose cyclic hemiacetals are the most easily made. The sugar molecules are in equilibrium between the open chain and cyclic structures. Details are given in Program 25.

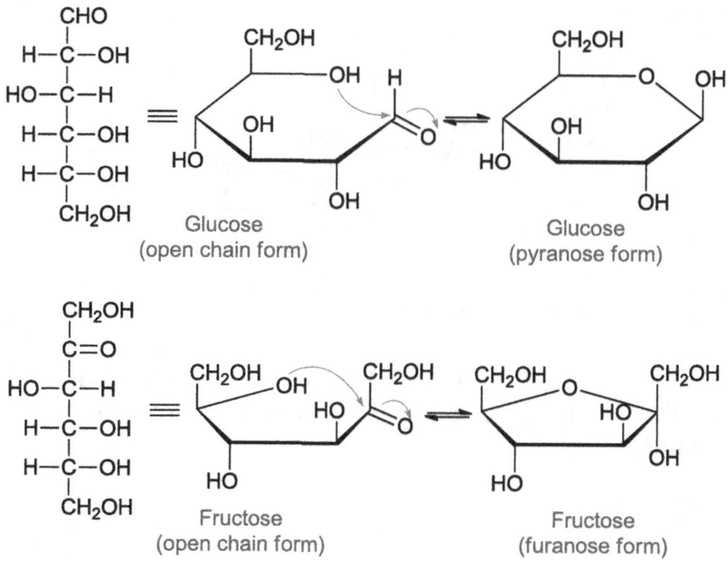

FIGURE 8.5
Cyclic hemiacetal formation (*Haworth projections*).

Hemiacetal formation gives a new chirality center which can have either of the two possible configurations. Because of the other fixed chirality centers in the sugar, two diastereomeric hemiacetals (anomers) are possible. As seen in Figure 8.6, these anomers are labeled α- or β- depending on the orientation of the –OH group at the anomeric carbon of the hemiacetal.

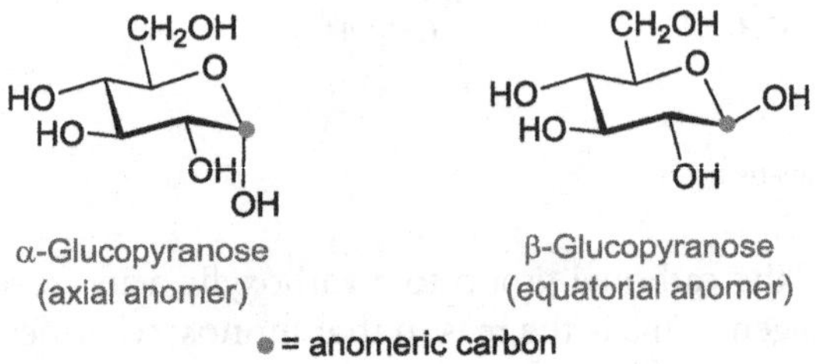

α-Glucopyranose
(axial anomer)

β-Glucopyranose
(equatorial anomer)

● = anomeric carbon

FIGURE 8.6
Pyranose chair conformations of glucose.

Another feature of simple sugars is their conformation. The pyranose structure of glucose can be compared to the cyclohexane system in Section 3.3.2. The favored chair forms of the hemiacetals in Figure 8.6 have different stabilities because of the orientation of the –OH substituent at the anomeric carbon.

In Section 7.7.1.4, we saw hemiacetals react with a second alcohol group to give acetals. Complex carbohydrates are polymers of simple sugars which are linked together as acetals. In carbohydrate chemistry, these are called glycosides. As Figure 8.7 shows, in glycosides, the only difference in the acetal formation is that the second alcohol function comes from a second sugar molecule (Figure 8.7).

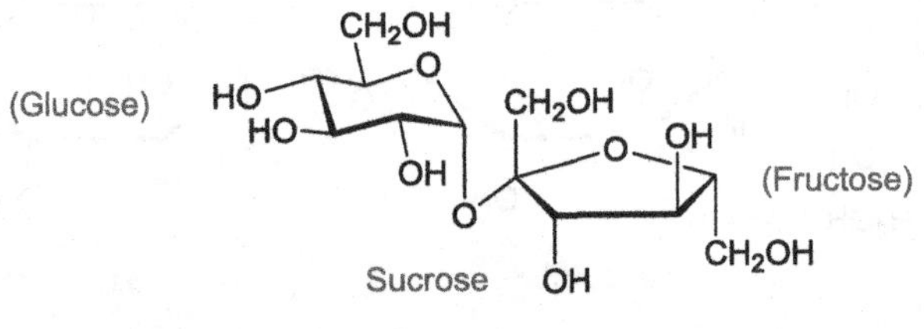

(Glucose)

(Fructose)

Sucrose

FIGURE 8.7
Sucrose glycoside (*acetal*) from glucose and fructose.

As Figure 8.8 shows, glycoside acetal bonds are easily broken by hydrolysis to give the simple sugar building blocks from the complex carbohydrate polysaccharides.

$$\text{Sucrose} \xrightarrow{\text{H}_3\text{O}^{\oplus}} \text{Glucose} + \text{Fructose}$$

$$\text{Cellulose} \xrightarrow{\text{H}_3\text{O}^{\oplus}} \text{Glucose monomers}$$

FIGURE 8.8
Hydrolysis of complex carbohydrates.

8.3   LIPIDS

The class name lipid refers to different compound types with various structures which have different functional groups. However, all lipids have large hydrocarbon content and are relatively non-polar. This makes them soluble in relatively non-polar organic solvents. It is this solubility, rather than a structural or chemical feature, that makes lipids different from other biomolecules.

As shown in Figure 8.9, lipids are classified as simple or complex. This depends on whether or not they have an ester group. The presence of an ester means they can undergo base hydrolysis.

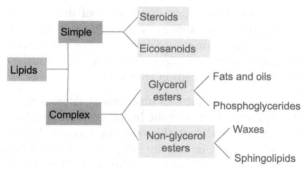

FIGURE 8.9
General classification of lipids.

Figure 8.10 shows two structurally different classes of simple lipids: steroids and eicosanoids. These have nothing in common except the physical property of similar solubility and no ester group. Other classes of simple lipids are terpenes and fat-soluble vitamins. These are not shown here.

Figure 8.10 also shows how complex lipids have the common feature of an ester group. They are further subdivided based on whether or not glycerol (1,2,3-propanetriol) is the alcohol portion of the ester.

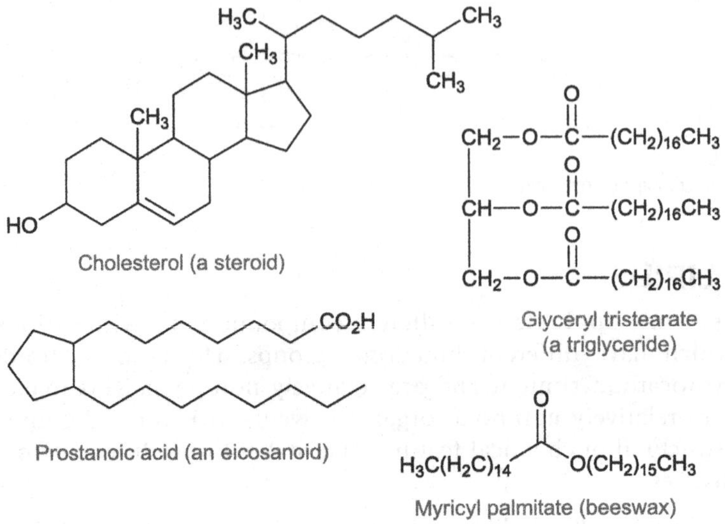

FIGURE 8.10
Examples of simple and complex lipids.

8.3.1 Fatty Acids

A discussion of lipids requires some knowledge of the fatty acids which are a major part of complex lipid structures. As Table 8.1 shows, fatty acids are carboxylic acids which have long unbranched hydrocarbon chains. Because of the biosynthetic routes that they come from, the acids usually contain even numbers of between 12 and 26 carbon atoms.

Fatty acids are classified as saturated or unsaturated, depending on the presence or absence of C=C double bonds in the hydrocarbon chain. Any C=C double bonds that are present, are almost always in the form of the *cis* stereoisomer. About 40 fatty acids occur in nature. In Table 8.1, palmitic and stearic are the most common examples for fully saturated fatty acids. Oleic and linoleic are the most common examples for unsaturated fatty acids.

Table 8.1 Selected Common Fatty Acids

Structural Formula	Common Name	IUPAC Name	M.P. (°C)
$CH_3(CH_2)_{10}CO_2H$	Lauric	Dodecanoic	43
$CH_3(CH_2)_{12}CO_2H$	Myristic	Tetradeconoic	55
$CH_3(CH_2)_{14}CO_2H$	Palmitic	Hexadecanoic	63
$CH_3(CH_2)_{16}CO_2H$	Stearic	Octadecanoic	70
$CH_3(CH_2)_7CH=CH(CH_2)_7CO_2H$	Oleic	*cis*-9-Octadecenoic	14
$CH_3(CH_2)_4(CH_2CH=CH)_2(CH_2)_7CO_2H$	Linoleic	*cis, cis*-9,12-Octadecadienoic	−5
$CH_3(CH_2CH=CH)_3(CH_2)_7CO_2H$	Linolenic	*cis, cis, cis*-9,12,15-Octadecatrienoic	−11

If there is only one C=C double bond, the fatty acid is monounsaturated. If there is more than one C=C double bond, the fatty acid is polyunsaturated. From Table 8.1, you can see that the melting point of fatty acids depends on both the length of the carbon chain and the number of C=C double bonds. The presence of *cis* double bonds strongly affects the efficiency of the hydrocarbon chains to pack together. This weakens the intermolecular attractions and lowers the melting point. The high percentage of non-polar hydrocarbon material means that fatty acids are not soluble in water and other polar solvents.

Except for the carboxylic acid group and C=C double bonds, fatty acids do not have any other functional groups. As a result, fatty acids undergo the reactions which are expected from these groups. Therefore, simple acid–base reactions, acyl conversions such as esterification and hydrolysis, and C=C additions such as halogenation and hydrogenation occur.

8.3.2   Fats and Oils

This is the biggest group of naturally occurring lipids. Fats usually come from animal sources and are solids at room temperature. Oils usually come from plants and are liquids at room temperature. Fats and oils are all triesters based on three molecules of a fatty acid, with glycerol as the common alcohol. Figure 8.11 shows esterification to give a triglyceride.

Simple triglycerides have only one type of fatty acid. Mixed triglycerides have more than one type of fatty acid. Again, the presence of C=C double bonds in the fatty acid gives products with lower melting points. Fats are usually solids because they have a high percentage of saturated fatty acids. Oils are usually liquids because they have a relatively higher percentage of unsaturated fatty acids.

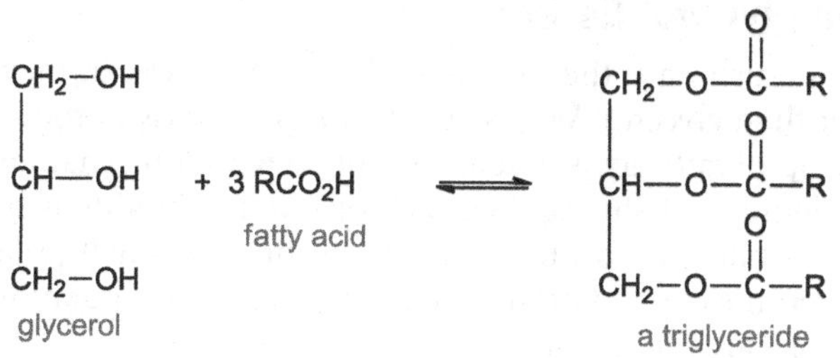

FIGURE 8.11
Triglycerides by esterification.

The functional group reactions of triglycerides are similar to those of the simple alkenes and esters seen earlier. For example, hydrogenation (Section 7.3.2) of unsaturated vegetable oils changes them to the more saturated solid or semi-solid fats. The most common reaction of the triglyceride ester groups is base hydrolysis to give soaps. This is often called saponification.

Soaps are generally mixtures of the carboxylate salts, usually sodium or potassium, of the fatty acids which come from the ester cleavage (Section 7.8.2). These salts work because of the difference between the ionic **hydrophilic** (water soluble) carboxylate and the non-polar **hydrophobic** (water insoluble) hydrocarbon ends of the molecule. These properties allow the soap to act as a cleaner by dissolving oil-soluble dirt into water.

Phosphoglycerides are closely related to triglycerides. In this important group of lipids, one of the fatty acids is replaced by a phosphoric acid group. This phosphoric group is further esterified with another alcohol. Figure 8.12 shows the two most important examples of the phosphoglycerides. These are the lecithins, which have choline as the other alcohol, and cephalins, which have ethanolamine as the other alcohol. The phosphoric ester gives the triglyceride molecule a polar hydrophilic charged head. The rest of the molecule is a relatively non-polar hydrophobic tail.

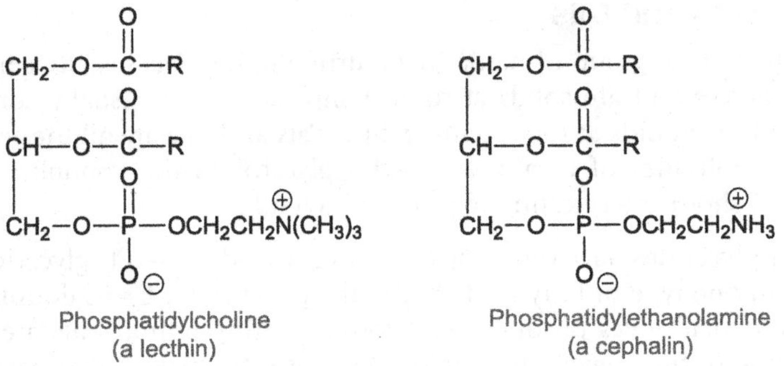

Phosphatidylcholine
(a lecthin)

Phosphatidylethanolamine
(a cephalin)

FIGURE 8.12
General structures of the lecthins and cephalins.

8.3.3 Non-glycerol Esters

These lipids can also have their esters hydrolyzed. However, they are esters of alcohols other than glycerol. Waxes are the simplest class of these compounds, and are made up of fatty acids that are esterified with long chain monohydroxy alcohols. As Figure 8.13 shows, their lack of water solubility is because of the large amount of aliphatic content relative to the very small polar ester head. Their melting points depend on the chain length and unsaturation of the acids and alcohols they are made from.

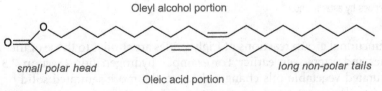

Oleyl alcohol portion

small polar head

long non-polar tails

Oleic acid portion

FIGURE 8.13
Typical structure of a wax.

Sphingolipids are another major class of non-glycerol esters. They have the amino dialcohol sphingosine instead of glycerol. All examples have a fatty acid linked to the amino group by an amide bond. They also have another acid component which is bonded to the 1° alcohol as an ester. If this is a phosphate ester, the products are the important subclass of sphingomyelins. See Figure 8.14.

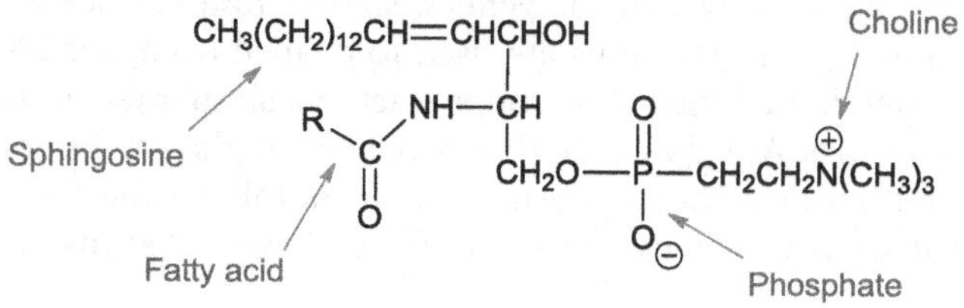

FIGURE 8.14
A typical sphingolipid (*Sphingomyelin*).

8.3.4 Eicosanoids

Members of the eicosanoid group are all related to the C_{20} fatty acid, eicosanoic acid, and its unsaturated equivalents. This includes prostaglandins, prostacyclins, thromboxanes, and leukotrienes. These compounds are found in all animal tissues. They have different, very specific controlling functions throughout the body.

Prostaglandins are a good example of the general class. Figure 8.15 shows that all prostaglandins come from arachadonic acid and have a 5-membered carbocyclic ring. Their class name is based on their first isolation from the prostate gland.

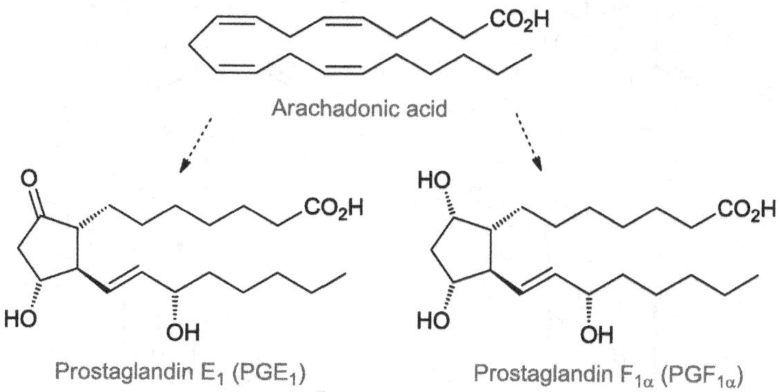

FIGURE 8.15
Selected prostaglandins from arachadonic acid.

Prostaglandin chemistry relates directly to the carboxylic acid, the C=C double bonds, and the hydroxy and/or carbonyl groups on the ring. All of these functional groups react in the same way as the simple molecules seen earlier. Also note the chirality centers and the C=C double bonds which allow for a number of possible stereoisomers.

8.3.5   Steroids

Steroids have the common structural feature of a tetracyclic system of three 6-membered and one 5-membered ring. Many different related structures are seen in plants, animals, and humans. Figure 8.10 shows cholesterol, the most common human example. Figure 8.16 shows that cholesterol is a starting material for the formation of other steroid hormones.

The examples below show that the general steroid framework stays the same. Simple functional group reactions are used to change from one compound to the next. In general, the functional groups react in a similar way to the reactions for simple molecules. As Figure 8.16 shows, oxidation plays an important role in many of these reactions. The tetracyclic framework follows the concepts of conformational analysis which was shown for simpler cyclic systems in Chapter 3.

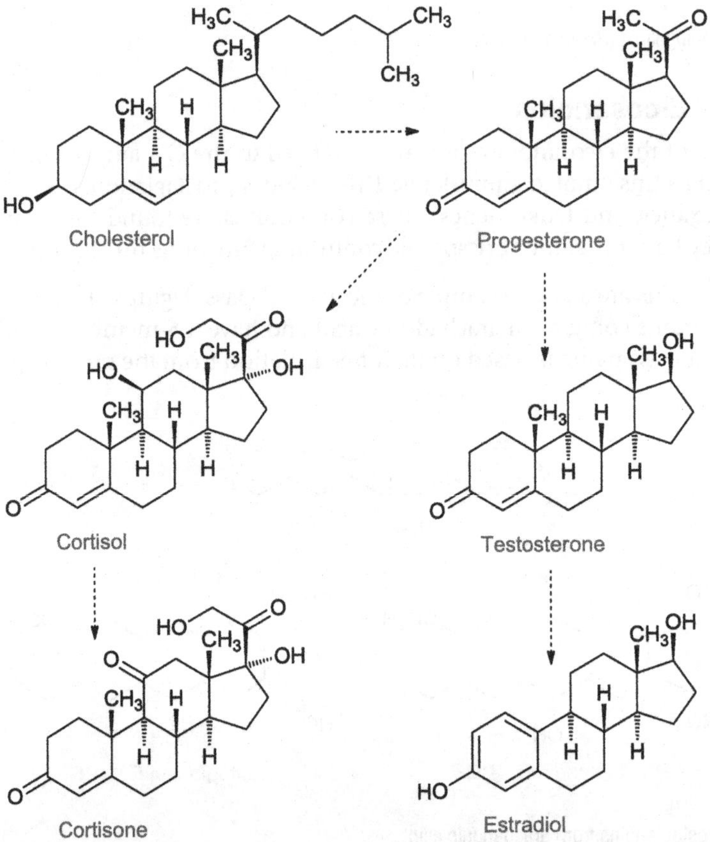

FIGURE 8.16
Cholesterol-derived steroidal hormones.

8.4   AMINO ACIDS, PEPTIDES, AND PROTEINS

Amino acids are compounds which have both amine and carboxylic acid functional groups. The 20 natural amino acids in Appendix 12 have the amine group on the α-carbon next to the carboxylic acid. The chemistry of amino acids is largely a combination of the reactions of the two individual functional groups. For example, the carboxylic group can be esterified by a mechanism discussed in Section 7.8.2. The amino group reacts with acyl components to give amides. This is the same as explained in Section 7.8.3. However, the physical properties of amino acids are more like salts rather than uncharged molecules. See Figure 8.17.

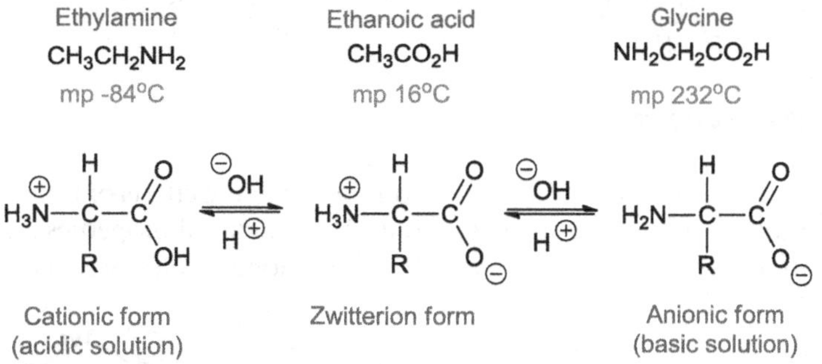

Ethylamine	Ethanoic acid	Glycine
$CH_3CH_2NH_2$	CH_3CO_2H	$NH_2CH_2CO_2H$
mp -84°C	mp 16°C	mp 232°C

Cationic form (acidic solution)	Zwitterion form	Anionic form (basic solution)

FIGURE 8.17
General α-amino acid zwitterion structure.

Amino acids are colorless solids with melting points >200 °C. They are usually soluble in aqueous medium, especially acidic or basic solutions. However, they show a lower acidity than simple carboxylic acids and a lower basicity than simple amines. In acidic medium the cation is the major species. In basic medium the anion is the major species.

For every amino acid there is an intermediate pH at which there is only the zwitterion. The zwitterion is a form in which there is both a cation and an anion. These charges cancel so that the zwitterion is overall neutral. This isoelectric point, pI, is a physical property of each amino acid and is the point at which the amino acid is least soluble.

Certain amino acids are essential components of all living cells. Peptides usually have less than 50 amino acids. Proteins are large polypeptides that have more than 50 amino acids. Both peptides and proteins can be seen as polymers of amino acids. In these, the amino acid units are joined by peptide amide bonds as seen in Figure 8.18.

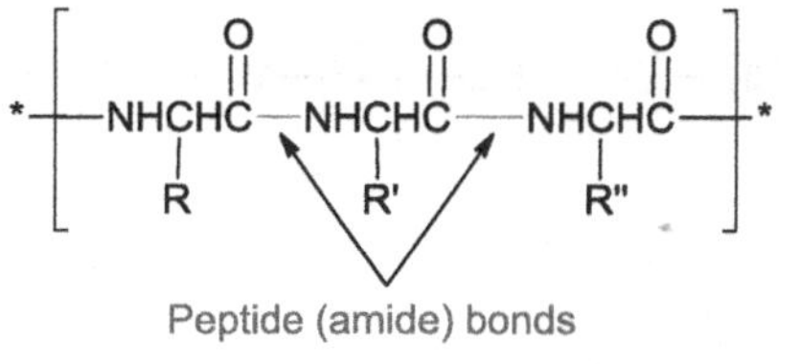

Peptide (amide) bonds

FIGURE 8.18
Typical peptide (*amide*) bonds between amino acids.

Amino acids can have a chirality center and stereoisomers can occur. However, all natural examples belong to the same stereochemical series and are in an optically pure state as only one of the enantiomers.

The peptide bond is the strongest and most important bond in proteins. The degree of double bond character of the C–N bond in amide was discussed in Section 7.8.3. Most peptides tend to favor the *trans* geometry shown in Figure 8.19.

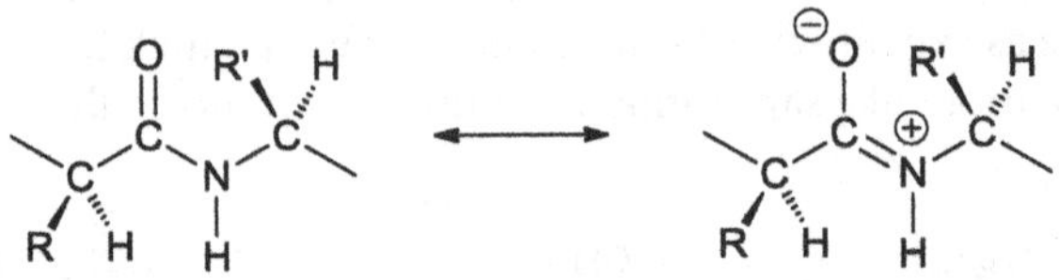

FIGURE 8.19
Trans geometry of the peptide bond.

The structural features of peptide bonds combine with the effects of intramolecular and intermolecular interactions. These control the chemical properties and overall conformational shapes which are shown by these biomolecules. See Figure 8.20.

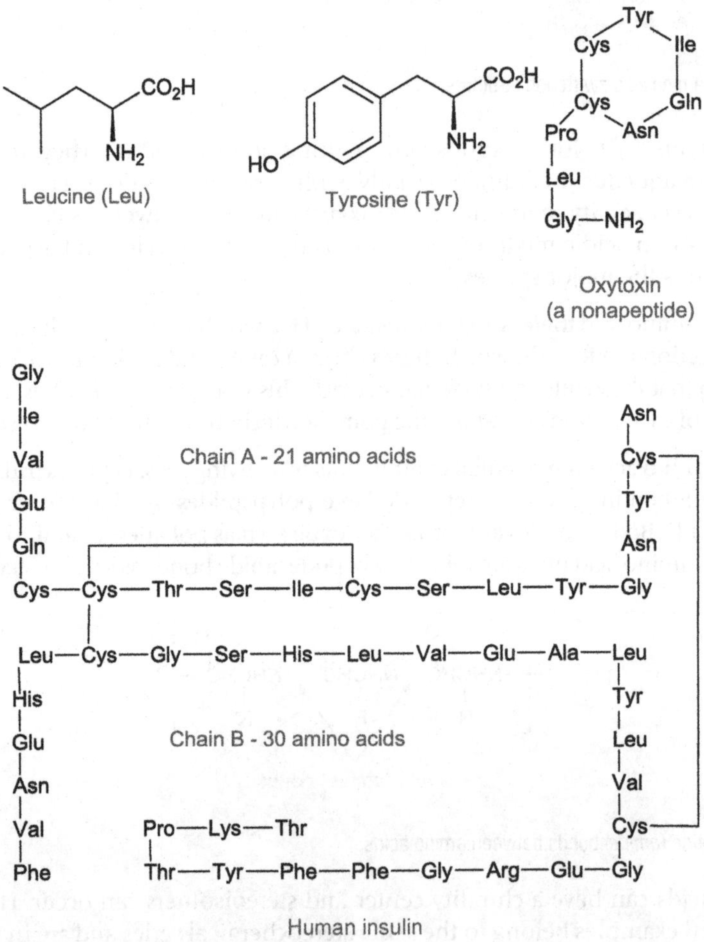

Leucine (Leu)

Tyrosine (Tyr)

Oxytoxin
(a nonapeptide)

Chain A - 21 amino acids

Chain B - 30 amino acids

Human insulin

FIGURE 8.20
Selected amino acids, a peptide, and a protein.

Proteins have four levels of structural complexity.

- Primary structure is the sequence in which the amino acids are connected.
- Secondary structure refers to the way pieces of the peptide structure are arranged because of intramolecular or intermolecular hydrogen bonding.
- Tertiary structure is the overall three-dimensional shape of the protein. This is the added result of all the secondary structures.
- Quaternary structure is the result of binding together of several protein molecules to give large structures which are the proteins in their functional role.

Overall, amino acids, peptides, and proteins have important structural and regulatory roles in all living cells.

8.5 NUCLEIC ACIDS

The nucleic acids, deoxyribonucleic acid (DNA) and ribonucleic acid (RNA), are the molecules responsible for carrying the genetic information of a cell. In the same way that proteins are polymers of amino acids, nucleic acids are long chain "polymers" of nucleotide building blocks. Each nucleotide is made up of a nucleoside along with phosphoric acid. As shown in Figure 8.21, each nucleoside is made up of a simple aldopentose sugar and a heterocyclic amine base. Base refers to the ability of the nitrogen lone pair to accept a proton.

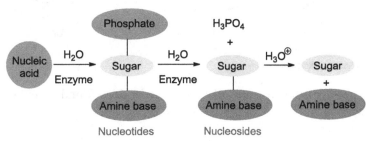

FIGURE 8.21
Nucleic acid composition.

Nucleic acids bring together the features of the sugars, phosphate esters, and amines which we studied earlier. Although nucleic acids seem complex, they follow much of the same simple functional group chemistry.

Figure 8.22 shows the only two sugars which occur in nucleic acids: β-D-ribose in RNA and β-D-2-deoxyribose in DNA These sugars differ from each other only in the presence or absence of the hydroxyl group at C_2. The naming of the nucleic acids is based on the names of these sugars.

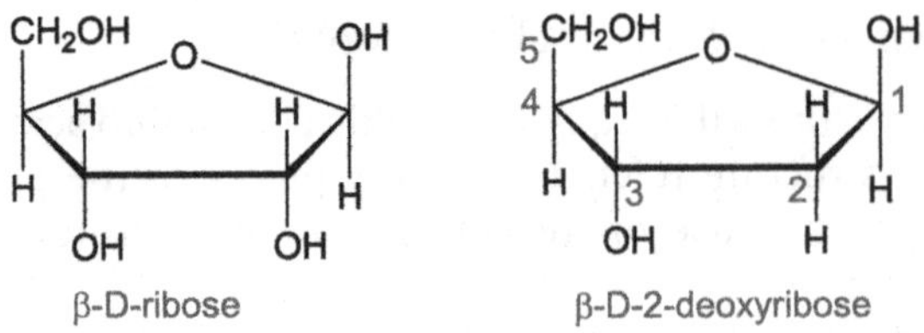

β-D-ribose

β-D-2-deoxyribose

FIGURE 8.22
The nucleic acid sugars.

Figure 8.23 gives the five heterocyclic nitrogen bases which occur in the nucleic acids. The two purines, adenine and guanine, are found in both DNA and RNA. Of the three pyrimidines, cytosine occurs in both DNA and RNA. Thymine is found only in DNA and uracil is only in RNA.

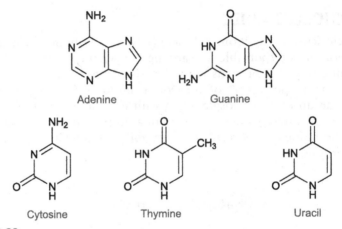

Adenine

Guanine

Cytosine

Thymine

Uracil

FIGURE 8.23
Nucleic acid heterocyclic bases.

Figure 8.24 gives the structure of typical nucleotides found in either of the nucleic acids.

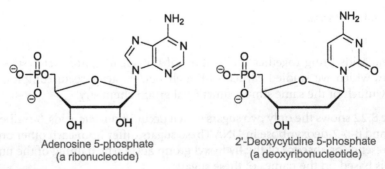

Adenosine 5-phosphate
(a ribonucleotide)

2'-Deoxycytidine 5-phosphate
(a deoxyribonucleotide)

FIGURE 8.24
Examples of nucleotides.

QUESTIONS AND PROGRAMS

Q 8.1. Classify the following simple sugars (monosaccharides).

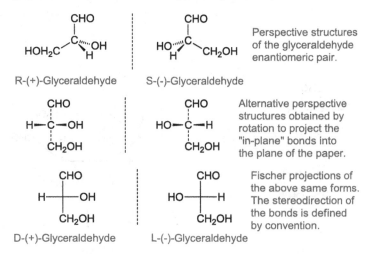

PROGRAM 25 Fischer/Haworth Diagrams

A It can be difficult at first to classify carbohydrates by their chirality centers. This is because there can be many chirality centers in the carbon backbone. This concept is simply an extension of stereoisomerism from Chapter 3. We now take a look at what this looks like for carbohydrate examples.

For single chirality centers you have seen perspective drawings. We now introduce the Fischer projection formula. This is best explained by an example.

R-(+)-Glyceraldehyde   S-(-)-Glyceraldehyde — Perspective structures of the glyceraldehyde enantiomeric pair.

— Alternative perspective structures obtained by rotation to project the "in-plane" bonds into the plane of the paper.

D-(+)-Glyceraldehyde   L-(-)-Glyceraldehyde — Fischer projections of the above same forms. The stereodirection of the bonds is defined by convention.

Try to show what the Fischer projection rules are.

B It should be clear that all vertical bonds project into the plane of the paper, and all horizontal bonds project out of the plane of the paper. In the case of carbohydrates the arrangement of the ligands is also important. The most oxidized ligand, either aldehyde or ketone, is drawn at the top of the vertical axis and the most reduced –CH$_2$OH end is at the bottom.

Continued...

—Cont'd

In Fischer's original assignment of the glyceraldehyde enantiomers, he labeled the dextrorotatory (+) isomer as D and the levorotatory (−) isomer as L. These labels are still used in carbohydrate chemistry today. The labels connect all carbohydrates to either the D- or L-series. In a Fischer diagram, this shows whether the bottom chiral center of the vertical chain has the same arrangement as D- or L-glyceraldehyde. The following example shows this.

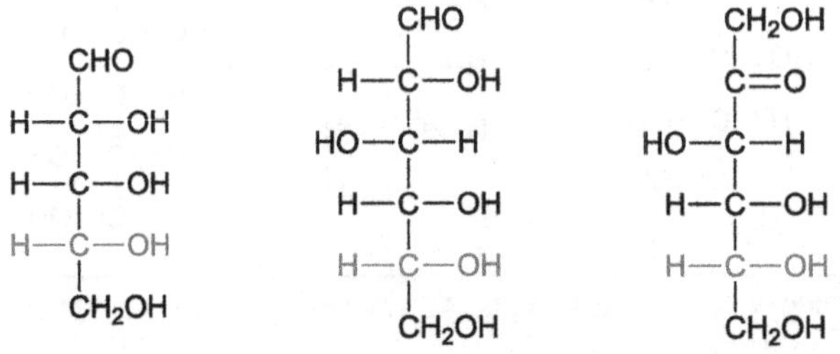

D-Ribose   D-Glucose   D-Fructose

All three have the same arrangement as D-glyceraldehyde with the OH group on the right. Therefore all are labeled as D-series sugars.

C   Because carbohydrates can have cyclic hemiacetal structures, the next problem is how to change from the Fischer projection to the cyclic Haworth projection. This projection is a view in which the ring is drawn as planar. The rings may be 5- or 6-membered, for furanose or pyranose, on which the substituents either point above or below the plane of the ring. The method for drawing a Haworth projection is shown below with the example of D-glucose.

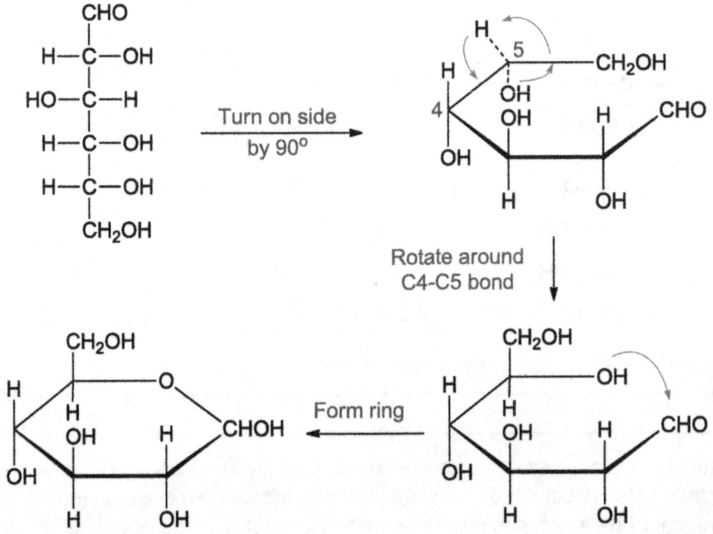

Turn on side by 90°

Rotate around C4-C5 bond

Form ring

D-Glucose
Haworth projection

Turn the Fischer projection on its side as shown. This causes the ligands on the right hand side to point below the proposed ring. The ligands from the left hand side point above the proposed ring.

For D-series sugars, you must rotate the C_4–C_5 bond as shown. This rotation places the C_5–OH in position for hemiacetal ring formation. As a result, the –CH_2OH group always points above the ring plane.

For L-series sugars, rotation in the opposite direction is needed. This rotation causes the –CH_2OH to point below the ring plane. Hemiacetal formation then gives the cyclic structure.

Clearly the hemiacetal formation can have either configuration at the new chirality center to give the α- (OH directed down) or β-anomer (OH directed up).

✍ Try to draw these Haworth projections.

D You should have drawn the following:

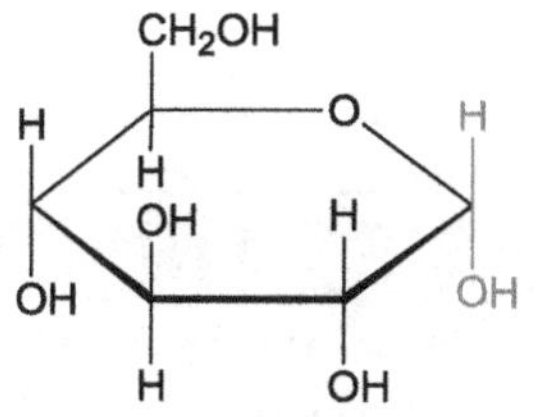

α-D-Glucose 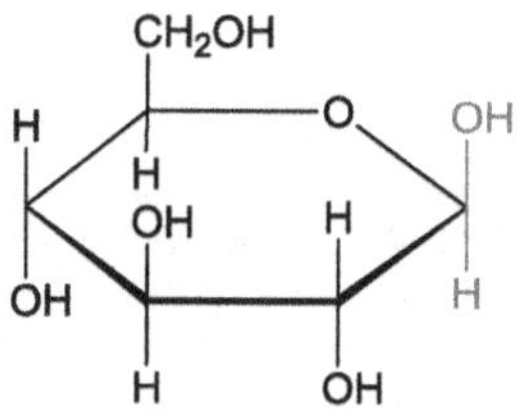 β-D-Glucose

---

Q 8.2. Draw the possible pyranose and furanose structures for D-fructose.

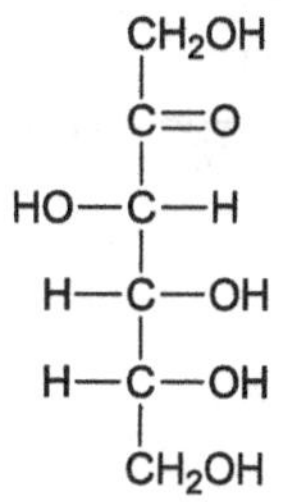

Q 8.3. Pick out any "reducing sugars" in **Q 8.1**.

Q 8.4. How many isomers, including stereoisomers, are possible in a triglyceride which is made from one unit each of palmitic, stearic, and oleic acids?

Q 8.5. The melting point of palmitoleic acid (*cis*-9-hexadecenoic acid) is −1 °C. This is very different from the melting point of palmitic acid which is 63 °C. Explain this.

Q 8.6. Hydrolysis of an optically active triglyceride gives one equivalent each of glycerol and oleic acid along with two equivalents of stearic acid. Draw a possible structural formula for the triglyceride.

Q 8.7. How many moles of hydrogen would be consumed during the hydrogenation of an oil which has one unit each of oleic, linoleic, and linolenic acids? Draw a structural representation of the product.

Q 8.8. Draw the general fused hydrocarbon ring system which is characteristic of a steroid. Explain why steroids are classified as lipids, although they do not have any fatty acid component.

Q 8.9. Draw the zwitterions, conjugate acids, and conjugate bases for the amino acids alanine and serine.

$$
\begin{array}{cc}
CO_2H & CO_2H \\
| & | \\
H_2N-C-H & H_2N-C-H \\
| & | \\
CH_3 & CH_2OH \\
\text{Alanine} & \text{Serine}
\end{array}
$$

PROGRAM 26 Amino Acid Isoelectric Points

A The isoelectric point (pI) is a physical property of each amino acid. At the isoelectric point, the solubility of the amino acid is minimized. This is because, at the isoelectric point pH, the concentration of the zwitterion is at a maximum. This can be shown by a study of the species that is formed from an amino acid as the pH changes.

$$
\overset{\oplus}{H_3N}-\overset{\overset{H}{|}}{\underset{\underset{CH_3}{|}}{C}}-CO_2H \quad \underset{H_3O^{\oplus}}{\overset{H_2O}{\rightleftharpoons}} \quad \overset{\oplus}{H_3N}-\overset{\overset{H}{|}}{\underset{\underset{CH_3}{|}}{C}}-CO_2^{\ominus} \qquad pK_a = 2.3
$$

$$
\overset{\oplus}{H_3N}-\overset{\overset{H}{|}}{\underset{\underset{CH_3}{|}}{C}}-CO_2^{\ominus} \quad \underset{H_3O^{\oplus}}{\overset{H_2O}{\rightleftharpoons}} \quad H_2N-\overset{\overset{H}{|}}{\underset{\underset{CH_3}{|}}{C}}-CO_2^{\ominus} \qquad pK_a = 9.9
$$

In the alanine example above, the zwitterion species dominates when the solution is between the pH values of 2.3 and 9.9.

✎ Comment on the species present at each of the pH extremes.

B   When the pH of a solution of an amino acid is equal to the pK_a of the acid, the concentration of the acid and its conjugate base are equal. At pH 2.3, the concentrations of the ammonium ion of alanine (conjugate acid) and the zwitterion are equal. At pH 9.9, the concentrations of the alanine carboxylate anion (conjugate base) and the zwitterion are equal.

The isoelectric point is the pH at which the concentration of the zwitterion is maximized. At higher pH values, the mixture of species moves toward an overall negative charge. At lower pH values, the move is toward an overall positive charge. For alanine, the isoelectric point is reached at pH 6.1.

Generally, each amino acid has a specific isoelectric point which is easily measured. The exact value of the isoelectric point is directly dependent on the structure of the amino acid. The effect of structure is greatest in amino acids which have side chain with acidic or basic groups. Proteins, polymers of amino acids, have isoelectric points which reflect the sum of their component amino acids.

The following material expands on the primary concepts in the book. Because they are not essential to the understanding of the concepts, they are collected here rather than distract from the material in the individual chapters.

APPENDIX 1: Electronegativity and Bond Polarity

Chemical bonds can be classed as non-polar covalent, polar covalent, or ionic based on the following:

1. Non-polar covalent: The electronegativity difference between the bonded atoms is in the range 0.0–0.4 units.
2. Polar covalent: The electronegativity difference between the bonded atoms is in the range 0.5–1.6 units.
3. Ionic: The electronegativity difference between the bonded atoms is 1.7 units or greater.

Average Electronegativities of Selected Elements

H 2.21							
Li 0.98			B 2.04	C 2.55	N 3.04	O 3.44	F 3.98
Na 0.93	Mg 1.31		Al 1.61	Si 1.90	P 2.19	S 2.58	Cl 3.16
K 0.82		Cu(I) 1.90	Zn 1.65				Br 2.96
							I 2.66
		Hg 2.00		Pb(IV) 2.33			

APPENDIX 2: Key IUPAC Rules for Substitutive Naming of Organic Compounds

1. The functional group for the basis of the name must be identified. See Appendix 3 for a partial list of functional group priorities. Usually, a functional group with carbon in the higher oxidation state with more bonds to heteroatoms takes precedence.

2. The longest continuous chain of carbon atoms which has the functional group is then the basis for the substitutive name. Atoms or groups other than hydrogen are treated as substituents.
3. The parent compound is named by adding the appropriate suffix to the name of the root hydrocarbon. The final -e in the hydrocarbon name is kept only if the suffix starts with a consonant.
4. Multiple carbon–carbon bonds can only be designated by suffixes. Except for compounds that have multiple carbon–carbon bonds in addition to the functional group, no parent compound name can have more than one systematic ending.
5. The parent chain is numbered so that the functional group has the lowest possible number. If the numbering can be done in more than one direction, it is done to give substituents the lowest set of numbers.
6. The number for the functional group in the parent compound is usually written before the hydrocarbon root name portion. Where the type of functional group involves the terminal carbon of the root chain (e.g.,: al for aldehydes), the number one is not added to the name.
7. The number for each substituent must appear in the name. It is usually written before the substituent name, and this is placed before the root hydrocarbon portion.
8. Numbers which occur together are separated by commas, and numbers and text are separated by hyphens.

APPENDIX 3: Substitutive Name Prefixes and Suffixes in Decreasing Order of Priority

Class	Group Formula	Prefix[a]	Suffix[b]
Carboxylic acid	–COOH	Carboxy	Carboxylic acid –oic acid
Sulfonic acid	–SO$_3$H	Sulfo	Sulfonic acid
Ester	–COOR	R Oxycarbonyl	R ...–carboxylate R ...–oate
Acid halide	–CO–Hal	Haloformyl Halocarbonyl	Carbonyl halide –oyl Halide
Amide	–CO–NH$_2$	Carbamoyl	Carboxamide –amide
Nitrile	–C≡N	Cyano	Carbonitrile –nitrile
Aldehyde	–CHO	Formyl Oxo	–al

Cont'd

Class	Group Formula	Prefix[a]	Suffix[b]
Ketone	$\diagdown C = O \diagup$	Oxo	–one
Alcohol/phenol	–OH	Hydroxy	–ol
Thiol	–SH	Mercapto	–thiol
Amine	$-NH_2$	Amino	–amine
Ether	–OR	R –oxy	–
Sulfide	–SR	R –thio	–

[a]The functional group is treated as a substituent.
[b]The functional group is part of the root compound and the suffix is added to the hydrocarbon stem.

APPENDIX 4: Further Amine Nomenclature

Primary amines are named by one-word substitutive names in one of two styles:

1. Based on the parent hydrocarbon to which $-NH_2$ is attached. The final -*e* in the parent is replaced by the systematic –amine ending. A number is used to give the position at which $-NH_2$ is attached. For example, $CH_3CH_2CH_2NH_2$ is 1-propanamine.
2. Based on the parent amine. For primary amines, this is ammonia, NH_3. The name of the attached group then changes amine in a one-word name. No number is needed since the group must be attached to nitrogen. For example, $CH_3CH_2CH_2NH_2$ is propylamine.

Style 1 is best for primary amines, but Style 2 is still used for amines of simple structure.

Symmetrical secondary and tertiary amines with simple identical groups attached to nitrogen are named by Style 2 with the inclusion of the appropriate multiplier (*di* or *tri*).

Unsymmetrical secondary and tertiary amines, in which the hydrocarbon groups are not the same, are named as *N*-substituted derivatives of the main primary amine. The largest or most complex primary amine parent is chosen. The other groups are treated as substituents with the positional identifier *N*. For example, $CH_3CH_2-NH-CH_3$ is named *N*-methylethanamine.

Quaternary Ammonium salts, $R_4N^\oplus$, are named as two-word names made up of the cation and anion components. The name of the cation is derived by alphabetically arranging the four groups bound to nitrogen ahead of the -ammonium suffix all as one word. This is then followed by the particular anion as a separate word. For example, $(CH_3CH_2)_2N^\oplus(CH_3)_2$ I⁻ is diethyldimethylammonium iodide.

APPENDIX 5: E, Z-Sequence Rules for Geometric Isomerism in Alkenes

A set of sequence rules gives priorities to the substituents on the carbons of a double bond. The groups on each carbon are independently given a priority order. If the higher priority groups are on the same side of the double bond, the alkene is designated Z (from the German zusammen, meaning together). If they are on opposite sides, the alkene is designated E (from the German entgegen, meaning opposite).

The condensed sequence rules are as follows:

1. Rank the atoms attached directly to the double bond carbons in order of decreasing atomic number.

(E)-2-Chloro-2-butene   (Z)-2-Chloro-2-butene

2. If a decision cannot be made based on the directly attached atoms, move to the second, third, and so on atom until a difference is found.

3. Multiple-bonded atoms are treated as having an equal number of single bonds.

APPENDIX 6: Cahn-Ingold-Prelog R/S Sequence Rules

This set of rules allows the assignment of a priority order to the four substituents attached to a chirality center. The rules are the same as above in Appendix 5, and only the application of the resulting priority order differs.

After determining the set of priorities, the stereochemical configuration is described as **R** (rectus, or "right") or **S** (sinister, or "left") as follows. The center to be assigned is viewed so that the substituent which has lowest priority four is pointing away behind the chiral center. The other three substituents now project like the spokes of a wheel. R or S is decided by the direction in which the substituent priorities go from 1→2→3.

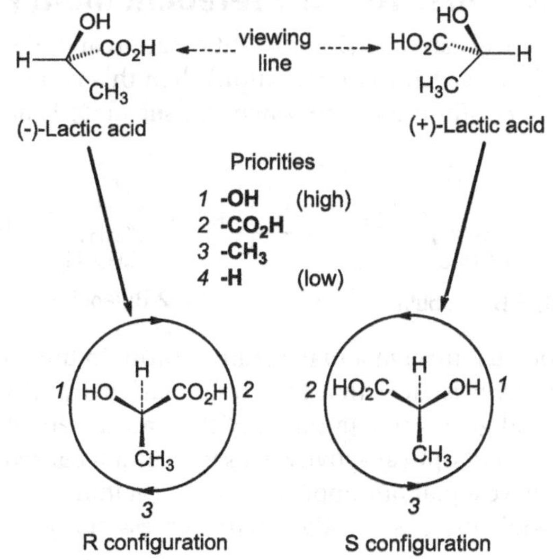

(-)-Lactic acid

(+)-Lactic acid

Priorities

1 -OH   (high)
2 -CO₂H
3 -CH₃
4 -H   (low)

R configuration

S configuration

APPENDIX 7: Selected Average Bond Energies (kJ/mol)

C—H	414	C—Cl	339	C—N	305
C—C	347	C—Br	285	C=N	464
C=C	620	C—I	213	C≡N	891
C≡C	837	C—S	272	Cl—Cl	243
C—O	360	C=S	536	Br—Br	192
C=O	741	O—H	464	I—I	151

APPENDIX 8: Syn- and Anti-Addition

Because a C=C lies in a plane, there are two sides or faces to the bond. Any stereochemical consequences of an addition reaction will be determined by whether the reagent components are added from the same face (*syn*) or opposite faces (*anti*) of the original C=C.

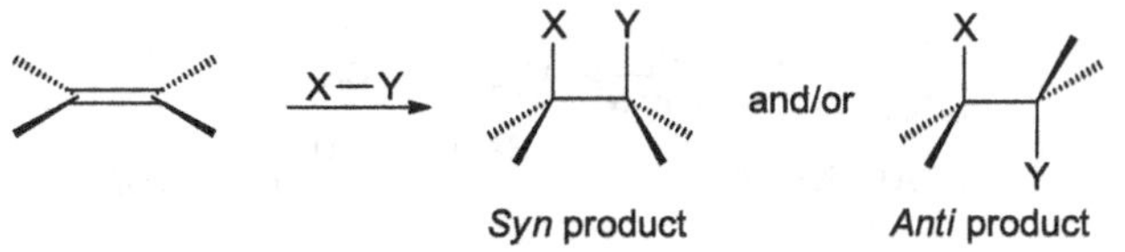

Syn product   *Anti* product

The hydrogenation of alkenes is an example. Because the process occurs on the surface of the catalyst, the reagent two hydrogen atoms are added from the same face to give the *syn*-addition product. The same outcome occurs with the hydroxylation of alkenes by permanganate. Both oxygen ligands are given to the same face of the alkene.

APPENDIX 9: Substitution Stereochemistry

S_N2 reactions occur with an inversion of configuration and, if the substrate center is chiral (i.e.,: has a specific configuration) then this is stereochemically significant. This is especially important when the substrate is an optically active enantiomer.

(S)-2-Bromobutane   (R)-2-Butanol

However, S_N1 proceeds through a planar carbocation. In this case, with an optically active substrate, it is clear that the configurational purity will be lost. In principle this should give equal amounts of the product enantiomers (racemic mixture) with no overall optical activity. This is a natural consequence of the fact that enantiomers have equal but opposite optical rotations. For reasons beyond the scope of the book, there is usually a partial excess of one enantiomer.

(S)-2-Bromobutane   (S)-2-Butanol

APPENDIX 10: Functional Group Preparations

The following is a summary of selected major preparations and their locations in the text.

Alkanes

Reduction (H_2/Pd or Ni) of alkenes or alkynes - catalytic hydrogenation (page 108–110).

UNSATURATED   SATURATED   Reduction

Alkenes

Dehydration of alcohols (page 115).

$$CH_3CH_2OH \xrightarrow[\text{Heat (170)}]{H_2SO_4} H_2C{=}CH_2$$

Elimination

Dehydrohalogenation of alkyl halides (page 114).

$$CH_3CH_2CH_2Br \xrightarrow[\text{Heat}]{\text{Base}} H_3CHC{=}CH_2$$

Elimination

Partial hydrogenation of alkynes (page 110).

$$RC{\equiv}CR' \xrightarrow[\text{H}_2\text{-Pd/CaCO}_3]{\text{Lindlar catalyst}} RHC{=}CHR'$$

Reduction

Alkynes

Alkyne carbanion nucleophile reactions (page 110–111).

$$RC{\equiv}CH \xrightarrow[\text{NH}_3\text{ (liq)}]{\text{Na}} RC{\equiv}C^{\ominus}Na^{\oplus} \xrightarrow{CH_3Br} RC{\equiv}CCH_3$$

Substitution

Double dehydrohalogenation of dihalides (page 114).

$$CH_3CHBrCH_2Br \xrightarrow[\text{Heat}]{\text{Base}} H_3CC{\equiv}CH$$

Elimination

Alkyl Halides

Electrophilic addition to alkenes (recall Markovnikov) (page 107–108).

$$H_3CHC{=}CH_2 \begin{array}{l} \xrightarrow{HBr} H_3CH_2C{-}CH_3 \\ \xrightarrow{Br_2} H_3CBrHC{-}CH_2Br \end{array}$$

Addition

Nucleophilic substitution of alcohols (page 115–116).

$$CH_3CH_2OH \xrightarrow[\text{PBr}_3\text{ or PBr}_5]{\text{conc. HBr}} CH_3CH_2Br$$

Substitution

Free radical halogenation of alkanes (page 104–105).

$$CH_3CH_3 \xrightarrow[\text{light (hv)}]{Cl_2} CH_3CH_2Cl$$

Substitution

Alcohols

Reduction of aldehydes, ketones, carboxylic acids, esters (page 119, 126–127).

RCHO / RCO$_2$H / RCO$_2$R' $\xrightarrow{\text{LiAlH}_4}$ RCH$_2$OH (primary)   *Reduction*

RCO.R'   RCH(OH)R' (secondary)

Hydration of alkenes (page 108).

H$_3$CHC=CH$_2$ $\xrightarrow[\text{H}_2\text{SO}_4]{\text{H}_2\text{O}}$ H$_3$CHĊ—CH$_3$ (OH)   *Addition*

Substitution of alkyl halides (page 112).

CH$_3$CH$_2$Cl $\xrightarrow{\ominus\text{OH}}$ CH$_3$CH$_2$OH   *Substitution*

Diazotisation of primary amines (page 128).

CH$_3$CH$_2$NH$_2$ $\xrightarrow{\text{HONO}}$ CH$_3$CH$_2$OH   *Substitution*

Ethers

Nucleophilic substitution of alkyl halides by alkoxides (page 114–115).

CH$_3$CH$_2$Br $\xrightarrow{\text{CH}_3\text{O}^{\ominus}\text{Na}^{\oplus}}$ CH$_3$CH$_2$OCH$_3$ *Substitution*

Carbonyl Compounds

Oxidation of alcohols (page 116).

CH$_3$CH$_2$OH (1°)   CH$_3$CHO (aldehyde)

$\xrightarrow[\text{H}^{\oplus}]{\text{K}_2\text{Cr}_2\text{O}_7}$ *Oxidation*

CH$_3$CH(OH)CH$_3$ (2°)   CH$_3$CO.CH$_3$ (ketone)

Oxidative cleavage of alkenes (page 109).

H$_3$CHC=CH$_2$ $\xrightarrow[\substack{\text{or}\\ \text{KMnO}_4/\text{H}^{\oplus}}]{\substack{\text{Ozonolysis}\\ \text{O}_3/\text{Zn-H}_2\text{O}}}$ CH$_3$CHO + CH$_2$O

May be oxidised   *Substitution*
further by KMnO$_4$

Hydration of alkynes (addition – tautomerism) (page 110).

$$RC{\equiv}CH \xrightarrow[\textit{dil. } H_2SO_4]{HgSO_4} \left[\begin{array}{c} H_3CC{=}CH_2 \\ | \\ OH \\ \text{enol} \end{array} \right] \rightleftharpoons \underset{\text{ketone}}{CH_3CO.CH_3}$$

Addition

Carboxylic Acids

Oxidation of 1° alcohols and aldehydes (page 116, 121–122).

$$CH_3CH_2OH \text{ / } CH_3CHO \xrightarrow[H^{\oplus}]{Na_2Cr_2O_7} CH_3CO_2H$$

Oxidation

Hydrolysis of nitriles (page 119, 124, 126).

$$H_3CH_2CC{\equiv}N \xrightarrow{H_3O^{\oplus}} CH_3CH_2CO_2H$$

Addition/Substitution

Hydrolysis of the acid derivatives, acid halides, esters, and amides (page 124–125).

$$\left. \begin{array}{c} CH_3COCl \\ CH_3CO_2R \\ CH_3CONHR \end{array} \right\} \xrightarrow{H_3O^{\oplus}} CH_3CO_2H \begin{array}{l} + \; HCl \\ + \; ROH \\ + \; NH_2R \end{array}$$

Substitution

Amines

Nucleophilic substitution of alkyl halides (page 127–128).

$$R{-}NH_2 \xrightarrow{R'X} R{-}\overset{H}{\underset{}{N}}{-}R'$$

Substitution

Reduction of nitriles, nitro compounds, amides (page 127, 130).

$$\begin{array}{c} RCH_2NO_2 \\ \text{or} \\ RCN \text{ / } RCO.NH_2 \end{array} \xrightarrow[\text{catalyst}]{H_2} RCH_2NH_2$$

Reduction

Amides

Nucleophilic substitution of acid derivatives (esters, acid halides) (page 125–126).

$$CH_3CO_2R \text{ / } CH_3CO.Cl \xrightarrow{R'NH_2} CH_3CO.NHR'$$

Substitution

APPENDIX 11: Functional Group Tests

The following are some of the major functional group diagnostic tests and their locations in the book. More modern approaches using spectroscopy have replaced many of these original chemical methods. However, at this stage, you will learn more from the chemical methods.

1. **Unsaturation** (alkenes, alkynes). Decoloration of Br_2 solution via addition, or of $KMnO_4$ solution via oxidative cleavage (page 108).
2. **Terminal unsaturation**. Evolution of CO_2 by complete oxidation after $KMnO_4$ oxidative cleavage to give $CH_2O \rightarrow HCO_2H \rightarrow CO_2$ (page 109).
3. **Methyl ketone group**. Haloform reaction (page 122).
4. **Carboxylic acid**. Acid–base reaction with $NaHCO_3$ to give off CO_2.
5. **Terminal alkyne**. Color reactions with $Cu(NH_3)_2^+$ (red organometallic) or $Ag(NH_3)_2^+$ (white organometallic) indicates terminal alkyne relatively acidic hydrogen (page 110).
6. **Alcohols**. Reaction with metallic Na to give off H_2 (acids will also respond in this test) (page 114).
7. **Amine classification**. Based on the diazotization reaction with nitrous acid (HNO_2). 1° amines leads to N_2 evolution, 2° amines give yellow oily layer, and 3° amines give no visible reaction (page 128).
8. **Alkyl halides**. Reaction with $AgNO_3$ gives a precipitate of silver halide (page ref. 111).
9. **Carbonyl compounds**. Aldehydes and ketones give orange/red precipitates of 2,4-dinitrophenylhydrazones with Brady's reagent (2,4-dinitro-phenyl-hydrazine) in a condensation reaction (page 121).
10. **Aldehyde**. Treatment with $Ag(NH_3)_2^+$ gives a silver mirror deposit (Tollens' test) based on the oxidation of the aldehyde by Ag^+ (page 121–122, 152).

APPENDIX 12: Most Common Amino Acids

Nonpolar R groups

Glycine (Gly)

Alanine (Ala)

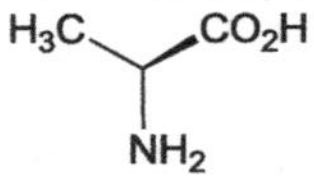

Valine (Val)

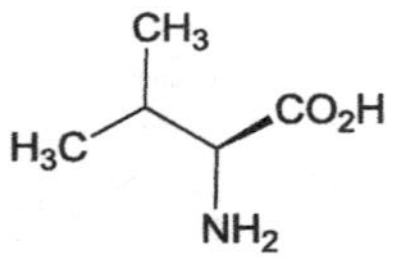

Leucine (Leu)

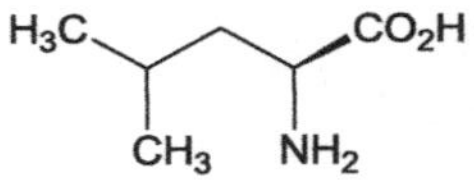

Isoleucine (Ile)

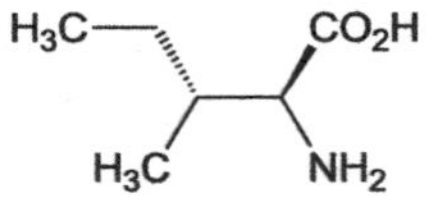

Methionine (Met)

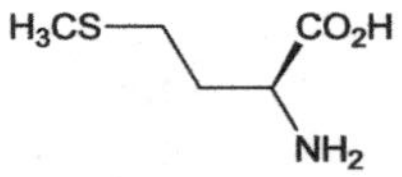

Phenylalanine (Phe)

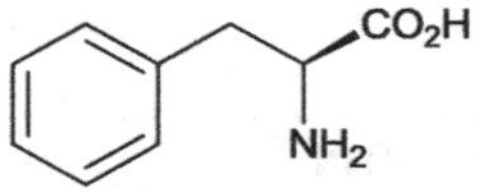

Proline (Pro)

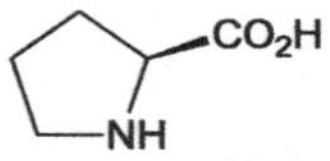

Polar neutral R groups

Serine (Ser)

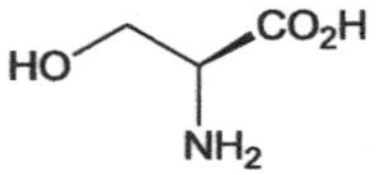

Threonine (Thr)

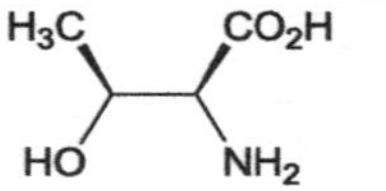

Cysteine (Cys)

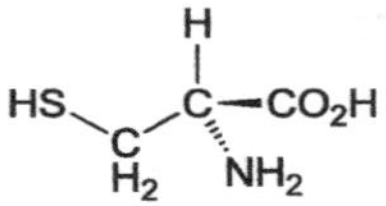

Asparagine (Asn)

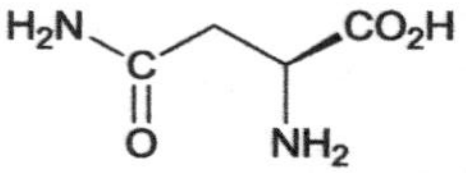

Tyrosine (Tyr)

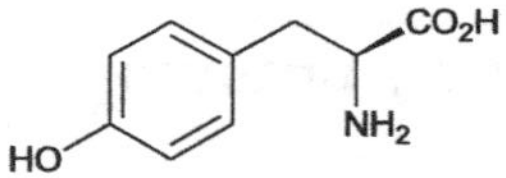

Tryptophan (Trp)

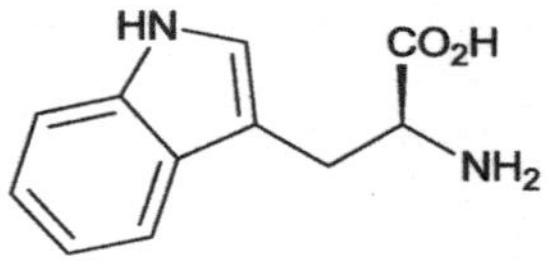

(Continued)

APPENDIX 12: Most Common Amino Acids—Cont'd

Nonpolar R groups

Glutamine (Gln)

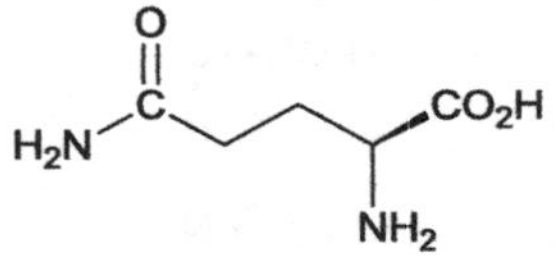

Acidic R groups

Glutamic acid (Glu)

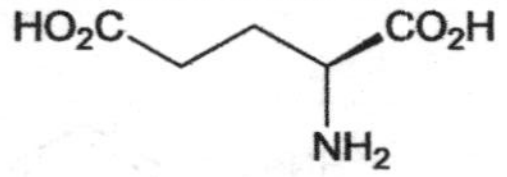

Aspartic acid (Asp)

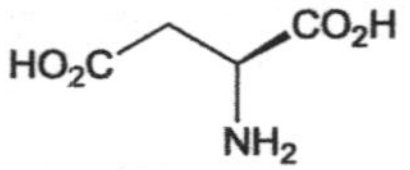

Basic R groups

Lysine (Lys)

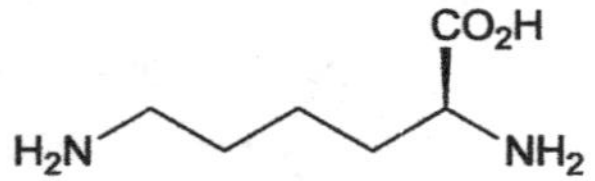

Histidine (His)

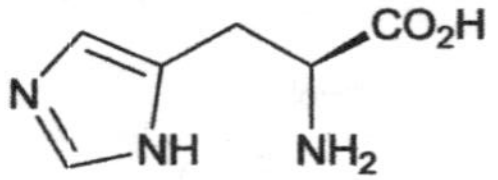

Arginine (Arg)

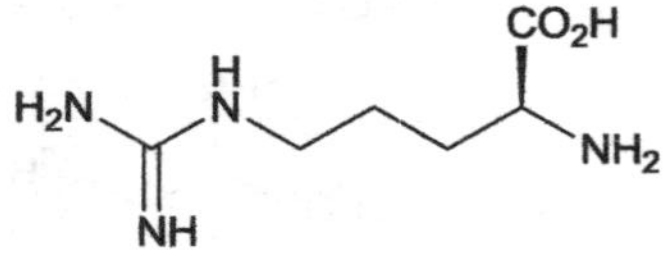

APPENDIX 13: Examples of Biological Significance

1. **Geometric Isomerism** (page 46–47)

Insect pheromones – Bombykol, is the compound that the female silkworm moth uses to attract a male. This compound has a specific trans/cis structure. The other three geometric isomers (*cis/cis, cis/trans, trans/trans*) do not work.

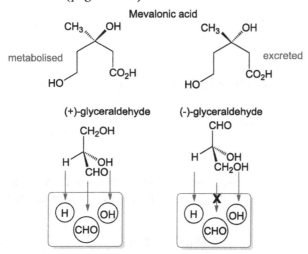

2. **Optical Isomers** (page 47–49)

Enzyme surface with specific binding sites

3. **Redox reactions** (page 18–19, 152)

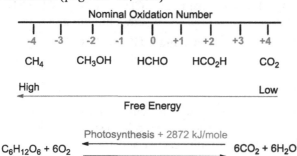

4. **Geometric Isomers** (page 46–47), **Alcohol Oxidation** (page 116), **and Carbonyl Condensation Reactions** (page 120–121)

The chemistry of sight depends on the formation and action of rhodopsin, and the vision process starts with the *cis-trans* isomerism to metarhodopsin.

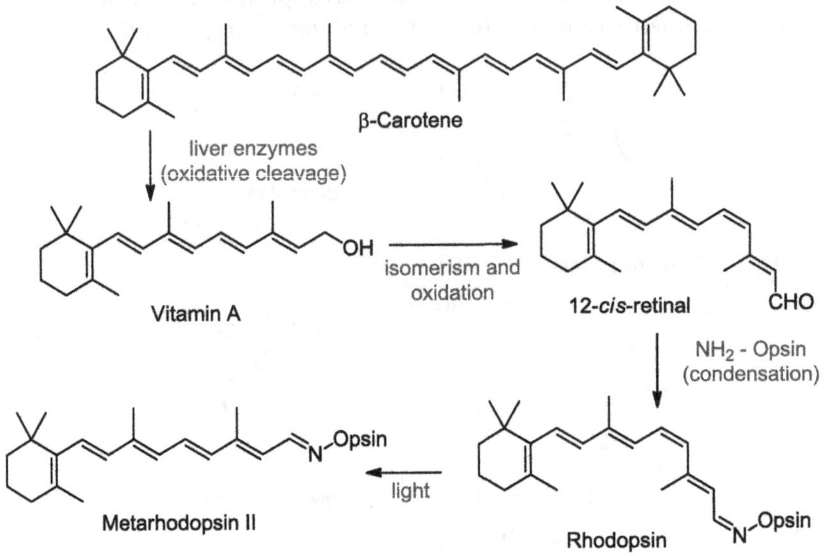

β-Carotene

liver enzymes (oxidative cleavage)

Vitamin A

isomerism and oxidation

12-*cis*-retinal   CHO

NH$_2$ - Opsin (condensation)

Metarhodopsin II

light

Rhodopsin

5. **Esters** (page 124–125)

There is a need for processed food to have a fresh smell and taste. This has led many esters being added as food flavor enhancers. Some of these are natural and others are not.

Ester	Flavor	Ester	Flavor
CH$_3$CH$_2$CH$_2$CO$_2$CH$_3$ methyl butanoate	Apple	CH$_3$CO$_2$(CH$_2$)$_7$CH$_3$ octyl acetate	Orange
CH$_3$(CH$_2$)$_2$CO$_2$(CH$_2$)$_4$CH$_3$ pentyl butanoate	Apricot	CH$_3$(CH$_2$)$_2$CO$_2$CH$_2$CH$_3$ ethyl butanoate	Pineapple
CH$_3$CO$_2$(CH$_2$)$_4$CH$_3$ pentyl acetate	Banana	HCO$_2$CH$_2$CH$_3$ ethyl methanoate	Rum

6. **Acyl substitution** (page 123–124)

Acyl interconversions are common in living systems. Acid chlorides and anhydrides are too reactive for reactions in cells. Esters are too unreactive. Living systems use thioesters as suitable reactive substrates.

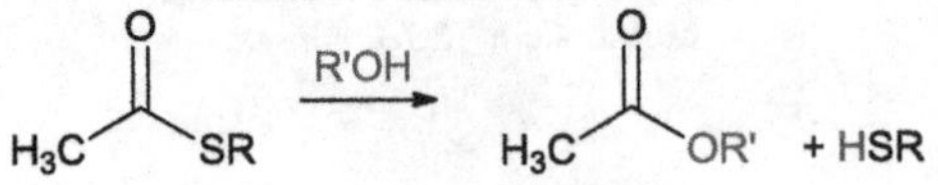

H$_3$C—C(=O)—SR →(R'OH) H$_3$C—C(=O)—OR' + HSR

The most important thioester is in acetyl coenzyme A. Its chemistry plays a vital role in the nervous system as an acetylation reagent for choline to give acetylcholine.

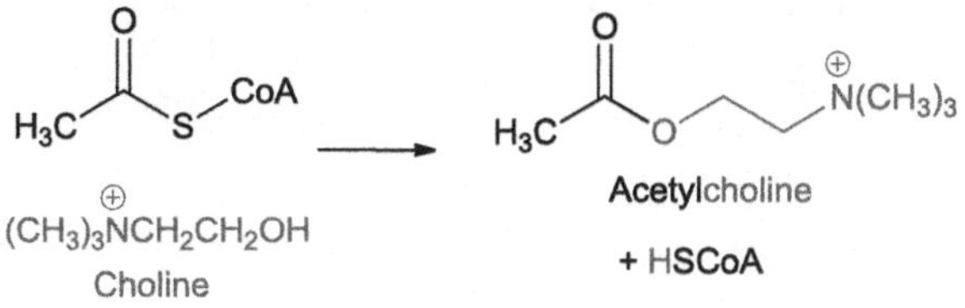

Acetylcholine

+ HSCoA

Choline

7. **Amines** (page 127–128)

Many nitrogen-containing compounds are physiologically active and affect the brain, spinal cord, and nervous system. These include the neurotransmitters dopamine, a lack of which causes Parkinson's disease; epinephrine to stimulate glucose production; and serotonin which is responsible for some nervous disorders.

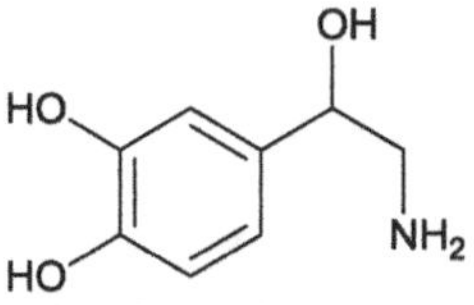

Dopamine

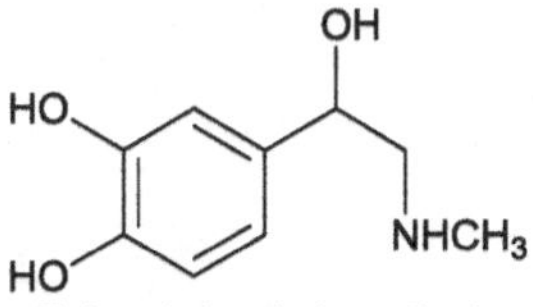

Epinephrine (adrenaline)

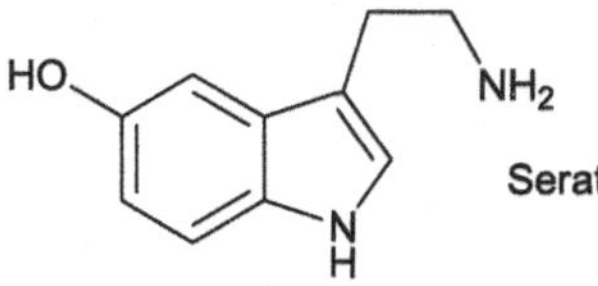

Seratonin

8. **Amides** (page 125–126)

Amides are well known because of the barbiturates and related designer drugs such as pentothal, barbital, 3-methyfentanyl. The last of these is highly addictive and about 3000 times more powerful than morphine.

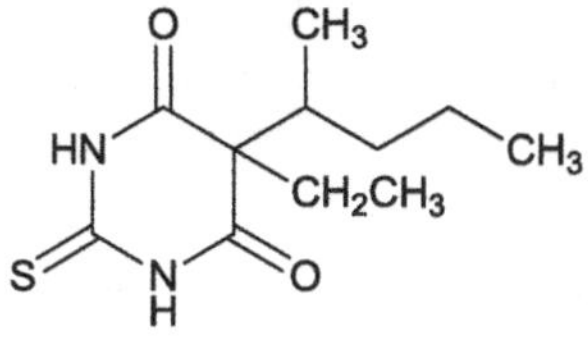

Pentothal

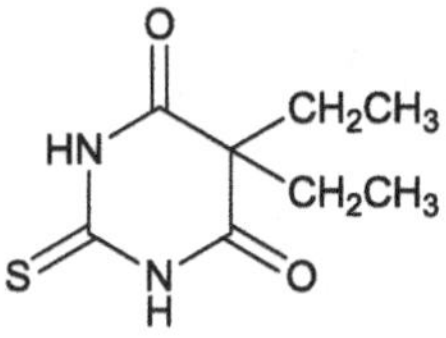

Barbital

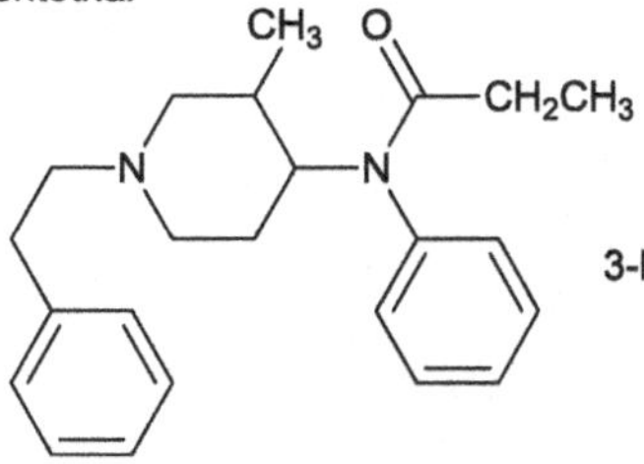

3-Methylfentanyl

Solutions

CHAPTER 1: ORGANIC STRUCTURES

Q 1.1.

$$H_3C{-}\overset{..}{\underset{..}{O}}{-}CH_3 \qquad CH_3\overset{..}{N}H_2 \qquad CH_3\overset{..}{\underset{..}{Br}}: \qquad H_3C{-}\overset{\overset{\displaystyle \overset{..}{\underset{..}{O}}}{\|}}{C}{-}\overset{..}{\underset{..}{O}}CH_3$$

Q 1.2. O, N, Br, O

Q 1.3.

 (a) C–H, N–H, O–H (b) B–H, C–H, O–H (c) C–S, C–N, C–O
 (d) C–I, C–H, C–Cl (e) B–H, C–N, C–F (f) C-B, C–Mg, C–Li

Q 1.4.

 (a) Alcohol (b) Aldehyde (c) Alkyne
 (d) Alkene (e) Carboxylic acid (f) Ketone

Q 1.5. The number and type of bonds that an atom forms shows the hybridization state. If only single σ bonds are involved, then sp^3 is assigned. If the atom is involved in one π bond, then sp^2 is assigned. Finally, if an atom is involved in two π bonds (*one triple or two double bonds*), then sp is assigned. When dealing with the hybridization states of noncarbon atoms, the presence of nonbonded lone pairs must be remembered.

$$\underset{sp^3 \;\; sp^3}{CH_3CH_2}\underset{sp^2}{\overset{\overset{\displaystyle O\; sp^2}{\|}}{C}}\underset{sp^3}{{-}OH}$$

$$\underset{sp^2}{CH_2}{=}\underset{sp^2}{\underset{\displaystyle |}{\overset{\displaystyle sp^3\; CH_3}{\overset{|}{\underset{sp^2}{C}{=}\overset{}{O}\; sp^2}}}}\underset{H}{}$$

$$\underset{sp}{HC}{\equiv}\underset{sp}{C}{-}\overset{sp^3\; H_2}{\underset{}{C}}{-}\underset{sp}{C}{\equiv}\underset{sp}{N}$$

Q 1.6.

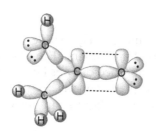

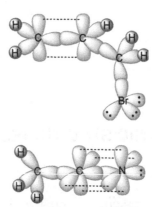

Q 1.7. The molecule has a nitrile and a ketone functional group, with hybridization states as shown. The sp^2 state of the ketone means that the two carbons bonded to it must be in the same plane as the C=O. The sp state of the nitrile must be linear. The molecule has 10 σ and 3 π bonds. The oxygen has two nonbonded pairs of electrons, and the nitrogen has one pair. The electronegative oxygen and nitrogen make the bonds that they are in very polar.

The ester has two oxygen heteroatoms each with two nonbonded pairs of electrons. The bonds with these electronegative atoms are polar. The sp^2 hybridization of the aromatic ring carbons means that they are all in the same plane. The molecule has 18 σ and 4 π bonds.

Study shows an amine, an alkyl halide, and a ketone. A total of 16 σ bonds and 1 π bond are present. The trigonal sp^2 of the C=O bond means that the two attached carbons lie in the same plane. Nonbonded pairs of electrons are present on the heteroatoms, N (*1 pair*), O (*2 pairs*), and Br (*3 pairs*). All of the bonds with the electronegative heteroatoms are polar.

Study shows ether, alkene, and alkyl halide functional groups. The sp^2 hybridization of the alkene means that this section of the molecule is in the same plane. Nonbonded pairs exist on both heteroatoms, O (2 pairs) and I (3 pairs). The bonds with the heteroatoms are polar. The overall bond count shows 12 σ bonds and 1 π bond.

Q 1.8.

C_4H_5NO, $C_8H_8O_2$, C_5H_8BrNO, C_4H_7IO

CHAPTER 2: FUNCTIONAL CLASSES I: STRUCTURE AND NAMING

Q 2.1.

(a) for example: HCO_2H, CH_3CO_2H, $CH_3CH_2CO_2H$, etc.

(b) for example: $HC\equiv CH$, $CH_3C\equiv CH$, $CH_3CH_2\equiv CH$, etc.

(c) for example:

Q 2.2. For example:

Q 2.3.

$C_4H_{10}O$ (a), (c), (f), (l) (identical)   (d), (i), (j) (identical)

(g), (h), (m) (identical), (e).

$C_3H_8O_2$ (b), (n)

C_3H_8O (k)

Q 2.4.

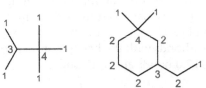

Q 2.5.

trivalent carbon

(a) $\overset{\downarrow}{C}H_2CH_2CH_3$

pentavalent carbons

(b) $CH_3\overset{\wedge}{CH}\!\!=\!\!CHCH_3$

hexavalent carbon

(c) $CH_3\!-\!\overset{\downarrow}{\underset{\underset{CH_2Cl}{|}}{C}}H_3\!-\!CH_3$

pentavalent carbon

(d)

trivalent oxygen

correct

(e) $CH_3\!-\!\underset{\underset{CH_2Cl}{|}}{CH}\!-\!CH_3$

(f) $\underset{CH_3}{\overset{CH_3}{\diagdown}}O\!-\!CH_3$

Q 2.6.

Q 2.7.

Q 2.8.

(a) 2,3-Dimethylhexane   (b) 4-Isopropyloctane   (c) 4-Ethyl-2,4-di-
                                                   methylhexane

(d) 3,3-Diethyl-1-   (e) 4-Methyl-2-hexene   (f) 4,5-Dimethyl-
pentene   1-hexyne

Q 2.9.

(a)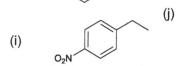

(b)

(c)

(d)

(e)

(f)

(g)

(h)

(i)

(j)

Q 2.10. *For example:*

(a) CH₃—C(CH₃)(CH₃)—CH₃ with NHCH₃   (b)   (c) CH₃CH₂CHO

(d) CH₃CH₂CH₂CH(CH₃)₂   (e) (CH₃)₂CHC(CH₃)₃

(f) H₂C=CHCN   (g) CH₃CHC(CH₃)₂ Br Br   (h)

(i) CH₃CH₂OCH₃   (j)

(k)   (l) CH₃CH=CHCH₃

Q 2.11.

(a) $H_3C-\overset{\overset{\displaystyle CH_3}{|}}{C}HOH$

2° alcohol

(b) $H_2C=CHCH_2OH$

1° alcohol

(c) $H_3C-\overset{\overset{\displaystyle CH_3}{|}}{C}HCH_2Br$

1° alkyl halide

(d) $(CH_3CH_2)_2CHCl$

2° alkyl halide

(e) $H_3CH_2C-\overset{\overset{\displaystyle CH_3}{|}}{\underset{\underset{\displaystyle CH_3}{|}}{N}}{}^{\oplus}-CH_2CH_3 \quad {}^{\ominus}OH$

4° ammonium salt

(f) $(CH_3)_2NH$

2° amine

(g) $H_3C-\overset{\overset{\displaystyle Br}{|}}{C}(CH_3)_2$

3° alkyl halide

(h) $H_3CC\overset{\nearrow O}{\underset{\searrow NHCH_3}{}}$

2° amide

(i) cyclopentane with CH_3 and OH

3° alcohol

Q 2.12.

(a) 2-Propanol

(b) 2-Propen-1-ol

(c) 1-Bromo-2-methylpropane

(d) 3-Chloropentane

(e) Diethyldimethylammonium hydroxide

(f) Dimethylamine

(g) 2-Bromo-2-methylpropane

(h) N-Methylethanamide

(i) 1-Methylcyclopentanol

Q 2.13.

Reason	Correct name
(a) Longest chain is pentane	2-Methylpentane
(b) Lowest numbering at C_2	2-Methylpentane
(c) Longest chain is butane	2,2-Dimethylbutane
(d) Longest chain is hexane	3,4-Dimethylhexane
(e) Lowest numbering is 3,3	3,3-Dimethylhexane
(f) Longest chain is heptane	4-methylheptane
(g) Longest chain is nonane	5-propylnonane
(h) Longest chain is hexane	2-methylhexane

Q 2.14.
(a) 3-Chlorobenzoic acid
(b) 3-Bromotoluene
(c) 4-Hydroxybenzenesulfonic acid
(d) 1-Phenyl-2-methylpropane

CHAPTER 3: ISOMERS AND STEREOCHEMISTRY

Q 3.1.

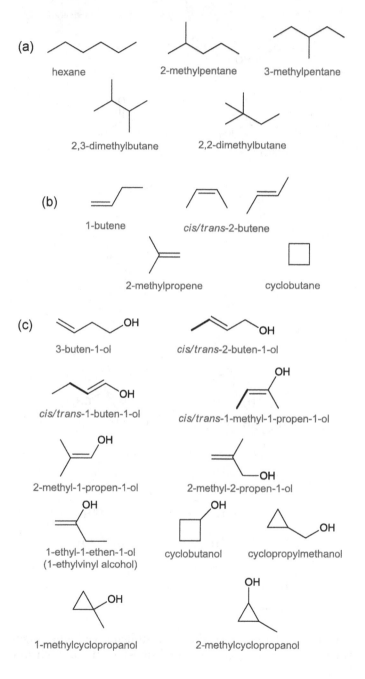

(a) hexane   2-methylpentane   3-methylpentane

2,3-dimethylbutane   2,2-dimethylbutane

(b) 1-butene   cis/trans-2-butene

2-methylpropene   cyclobutane

(c) 3-buten-1-ol   cis/trans-2-buten-1-ol

cis/trans-1-buten-1-ol   cis/trans-1-methyl-1-propen-1-ol

2-methyl-1-propen-1-ol   2-methyl-2-propen-1-ol

1-ethyl-1-ethen-1-ol (1-ethylvinyl alcohol)   cyclobutanol   cyclopropylmethanol

1-methylcyclopropanol   2-methylcyclopropanol

(d)   3-butenal   *cis/trans*-2-butenal   2-methylpropenal

cyclopropanecarboxaldehyde

(e)   Butanamide   2-methylpropanamide

N-methylpropanamide   *N*-ethylethanamide

N-propylmethanamide   *N*-isopropylmethanamide

Q 3.2.

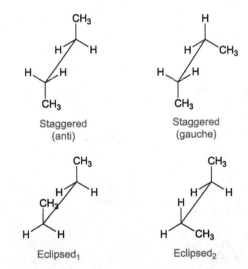

Staggered
(anti)

Staggered
(gauche)

Eclipsed₁

Eclipsed₂

Q 3.3. The staggered (*anti*), since the substituents with the largest van der Waals radii (*the CH₃ groups*), are furthest apart and their interaction is lowest.

Q 3.4. Only (c) may exhibit *cis/trans* stereoisomerism.

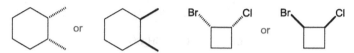

Cis-4-methyl-2-pentene   Trans-4-methyl-2-pentene

Q 3.5. (b) and (d) may exhibit *cis/trans* stereoisomerism.

Q 3.6. Only (a) and (h) can exhibit *cis/trans* stereoisomerism.

Q 3.7.

(a) The formula gives a DBE = 2. This result, and the three methyl groups, can be satisfied by the following triple bond, two double bonds, or a cyclopropene ring

$HC\equiv CC(CH_3)_3$ $CH_3C\equiv CCH(CH_3)_2$ alkynes

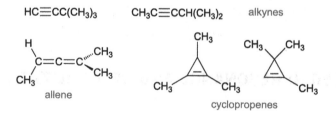

allene   cyclopropenes

(b) The formula indicates a DBE = 1. This result, together with an ethyl group and no stereoisomerism, can be satisfied by either a terminal double bond or a cyclopropane ring.

(c) The DBE = 1 is used by the aldehyde. The presence of a 2° alcohol, gives only one possible structure.

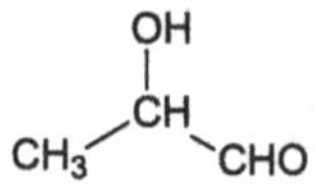

$$CH_3-\underset{\underset{OH}{\overset{\overset{CH}{|}}{|}}}{CH}-CHO$$

Q 3.8.

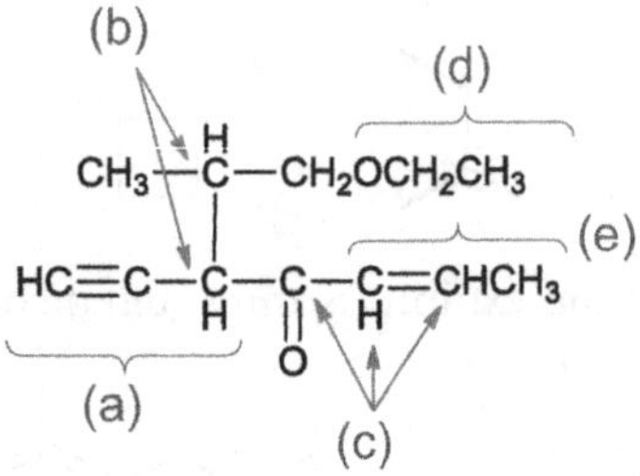

(b)

(d)

$$CH_3-\overset{\overset{H}{|}}{\underset{|}{C}}-CH_2OCH_2CH_3$$

(e)

$$HC\equiv C-\overset{|}{\underset{H}{C}}-\overset{\overset{|}{C}}{\underset{\overset{\|}{O}}{C}}-\overset{|}{\underset{H}{C}}=CHCH_3$$

(a)

(c)

Q 3.9.

Nootkatone
(grapefruit oil)

Camphor

Cholesterol

Nicotine

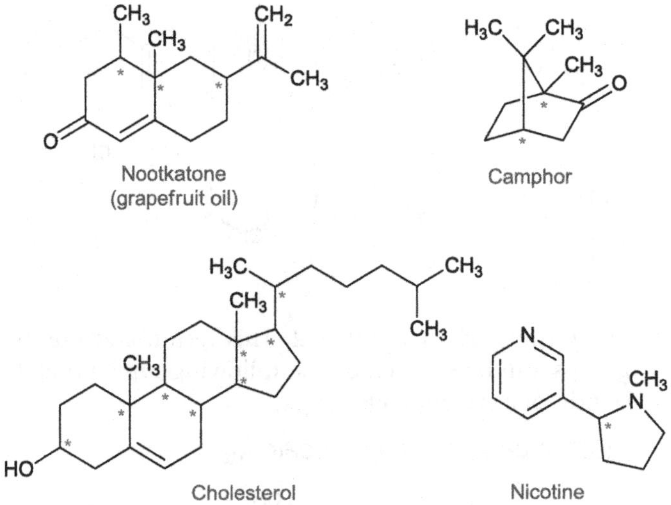

CHAPTER 4: RESONANCE AND DELOCALIZATION

Q 4.1.

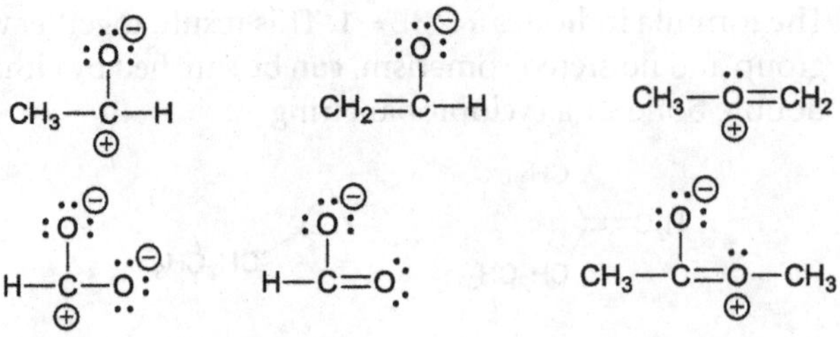

$$CH_3-\overset{\overset{:\overset{\ominus}{O}:}{\|}}{\underset{\oplus}{C}}-H$$

$$CH_2=\overset{\overset{:\overset{\ominus}{O}:}{\|}}{C}-H$$

$$CH_3-\overset{\overset{..}{O}}{\underset{\oplus}{=}}CH_2$$

$$H-\overset{\overset{:\overset{\ominus}{O}:}{\|}}{\underset{\oplus}{C}}-\overset{..}{\overset{..}{O}}:^{\ominus}$$

$$H-\overset{\overset{:\overset{\ominus}{O}:}{\|}}{C}=\overset{..}{O}:$$

$$CH_3-\overset{\overset{:\overset{\ominus}{O}:}{\|}}{\underset{\oplus}{C}}=\overset{..}{O}-CH_3$$

Q 4.2.

Q 4.3.

In the two aromatic examples above, the alternative initial resonance form for each structure is not shown. These do, however, represent important resonance structures and contribute greatly to the overall stability of the aromatic systems.

CHAPTER 5: REACTIVITY: HOW AND WHY

Q 5.1.

Q 5.2.

Ionic electrophiles	$H^{\oplus}$　　$Br^{\oplus}$　　$(CH_3)_3C^{\oplus}$
Non-ionic electrophiles	$\begin{array}{c}H\\\diagdown\\C=O\\\diagup\\H\end{array}$　　$CH_3-C\equiv N$
Ionic nucleophiles	$^{\ominus}OH$
Non-ionic nucleophiles	CH_3-OH　　NH_3　　$CH_3-C\equiv C-H$

Q 5.3. Because opposite charges attract, the feature of all polar reactions is the interaction of electron-rich nucleophilic sites in one molecule with electron-poor electrophilic sites in another. Bonds are formed if the nucleophile donates a pair of electrons to the electrophile, and bonds are broken if one product fragment leaves with an electron pair.

Q 5.4.

(a) Addition　　(b) Addition　　(c) Elimination　　(d) Substitution
(e) Addition　　(f) Substitution　　(g) Elimination

Q 5.5.

(a)　$CH_3-\overset{H}{\underset{\oplus}{C}}-CH_3$　　2° carbocation with greater inductive stabilization

(b)　$CH_3-\overset{H}{\underset{\oplus}{C}}-\overset{H}{C}=CH_2$　⟷　$CH_3-\overset{H}{C}=\overset{H}{C}-\underset{\oplus}{C}H_2$

Carbocation with resonance stabilization

(c)　⟷　etc

Oxyanion (alcoholate) with resonance stabilization

(d) $(CH_3)_3C$ ·   3° radical with greater
                       inductive stabilization

(e)

The carbanion with a resonance form which
delocalizes the negative charge onto oxygen

(f)   2° carbocation with greater inductive
                           stabilization. Note that the alternative
                           1° is not conjugated and cannot be
                           resonance stabilized.

Q 5.6. Since both reactions are shown as one-step processes, no reaction intermediates are involved. Suitable transition states can be drawn as follows:

Substitution   Elimination

Q 5.7. For the reactions to proceed without the participation of the hydroxide
(*either as nucleophile for substitution or as base for elimination*) means that
the C–Br bond must break by spontaneous heterolysis (*ionization*).
The transition state and intermediate for this process can be repre-
sented as

Transition state   Carbocation
                          intermediate

The overall reaction is controlled by one of two pathways: arrival of
the nucleophile to give substitution or the loss of a proton to give
elimination. The transition states for these processes can be drawn as
follows:

Alternative possible transition states

Q 5.8. You should have identified the following.

(e) $CH_3C\equiv CH$ $\xrightarrow{H_2/catalyst}$ $CH_3CH=CH_2$
     $\;\;0\;\;\;\;-1$ $\qquad\qquad\qquad\qquad\qquad -1\;\;\;\;-2$

Reduction with an overall change of -2

(g) $CH_3CH_2\underset{\underset{OH}{|}}{\overset{0}{C}}HCH_3$ $\xrightarrow{oxidant}$ $CH_3CH_2\underset{\underset{O}{\|}}{\overset{+2}{C}}CH_3$

Oxidation with an overall change of +2

CHAPTER 6: ACIDS AND BASES

Q 6.1.

Acid	Conjugate base	Base	Conjugate acid
$H_3O^{\oplus}$	H_2O	–	–
H_2O	$HO^{\ominus}$	H_2O	$H_3O^{\oplus}$
–	–	$CH_3S^{\ominus}$	CH_3SH
NH_3	$^{\ominus}NH_2$	NH_3	$^{\oplus}NH_4$
CH_3OH	$CH_3O^{\ominus}$	CH_3OH	$CH_3OH_2^{\oplus}$
CH_3NH_2	$CH_3NH^{\ominus}$	CH_3NH_2	$CH_3NH_3^{\oplus}$

Q 6.2.

(a) $CH_3\overset{..}{\underset{..}{O}}H$ $\quad H-\overset{\oplus}{N}H_3$ $\longrightarrow$ $CH_3\overset{..}{O}H_2^{\oplus}$ $+$ $:NH_3$

    Base   Acid   Conjugate acid   Conjugate base

(b) CH_3O-H $\quad \overset{\ominus}{:}NH_2$ $\longrightarrow$ $CH_3\overset{..}{\underset{..}{O}}:^{\ominus}$ $+$ $:NH_3$

    Acid   Base   Conjugate base   Conjugate acid

(c) $(CH_3)_2C=\overset{..}{\underset{..}{O}}:$ $\quad H-\overset{\oplus}{O}H_2$ $\longrightarrow$ $(CH_3)_2C=\overset{\oplus}{O}H$ $+\; H_2O$

    Base   Acid   Conjugate acid   Conjugate base

Q 6.3. The amide ion is the stronger base and the favored reaction is:

$$H_2N^{\ominus} + H_2O \longrightarrow H_3N + HO^{\ominus}$$

The conclusion is that H_2O (pK_a 15.5) is a stronger acid than NH_3 (pK_a 33).

Q 6.4. Since acetone has the lower pK_a, it is a stronger acid than NH_3 and the reaction will proceed as written.

Q 6.5. Since carbonic acid has the lower pK_a, it is a stronger acid than methanol. This means that the methoxide conjugate base is stronger than the bicarbonate anion and only the reverse reaction is favored.

Q 6.6. The correct order is as shown in the question.

$$CH_3CO.CH_2CO.CH_3 > CH_3CO.CH_2CO_2CH_3 > CH_2(CO_2CH_3)_2$$

The respective pK_a values are 8.8, 11, and 13.5. This agrees with the resonance stabilization that is possible. Since the ester group is less efficient at stabilization because of competing resonance forms, the more esters that are introduced, the less the overall resonance stabilization (see Program 17).

Q 6.7. The correct resonance forms are

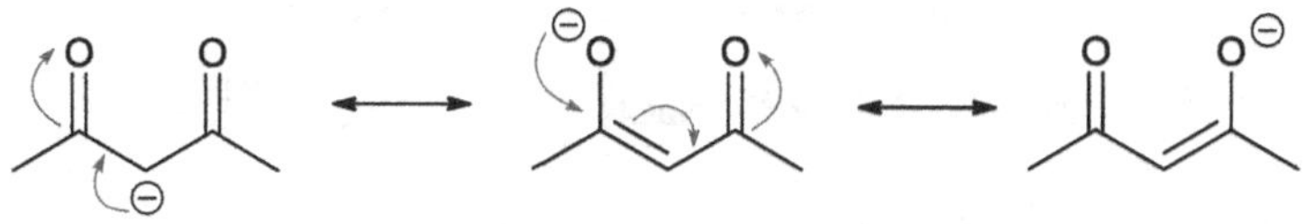

Q 6.8.

$CH_3CH_3 < CH_3CH_2OH < CH_3CHO < Cl_3CH_2CH_2CO_2H \leq CH_3CO_2H$
$< ClCH_2CO_2H$

CHAPTER 7: FUNCTIONAL CLASSES II: REACTIONS

Q 7.1.

(a) Electrophilic addition

$$\underset{\text{Nucleophile}}{CH_3CH{=}CH_2} \xrightarrow[Br^{\ominus}\ \text{Nucleophile}]{Br^{\oplus}\ \text{Electrophile}} CH_3CH(Br)CH_2Br$$

(b) Nucleophilic addition

$$CH_3CHO \xrightarrow[H^{\oplus}\ \text{Electrophile}]{^{\ominus}CN\ \text{Nucleophile}} CH_3CH(CN)OH$$

(c) Elimination

$$CH_3CH_2CH(Br)CH_3 \xrightarrow{^{\ominus}OH\ \text{Base}} CH_3CH_2CH{=}CH_2$$

(d) Nucleophilic substitution

$$CH_3CH_2CH_2Br \xrightarrow[\text{Nucleophile}]{CH_3C\equiv C^{\ominus}} CH_3CH_2CH_2C\equiv CCH_3$$

Electrophile

(e) Electrophilic addition

$$CH_3C\equiv CH \xrightarrow[\underset{\text{Nucleophile}}{Br^{\ominus}}]{\overset{H^{\oplus} \text{ Electrophile}}{}} CH_3C(Br)=CH_2$$

Nucleophile

(f) Nucleophilic acyl substitution

$$CH_3CH_2CO_2CH_3 \xrightarrow[\text{Nucleophile}]{CH_3NH_2} CH_3CH_2CO.NHCH_3$$

Electrophile

Q 7.2.

(a)

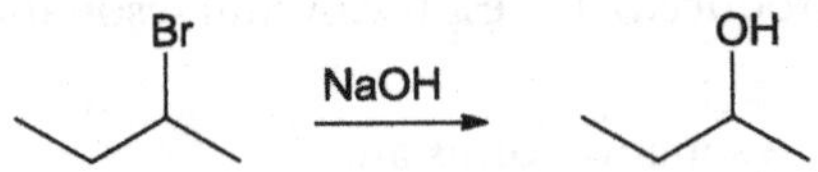

Alkyl halide   NaOH   Alcohol

(b)

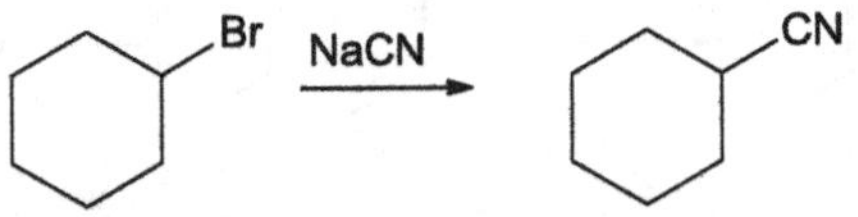

Alkyl halide   NaCN   Nitrile

(c)

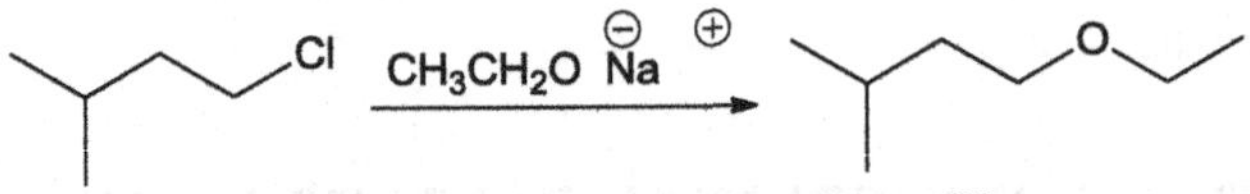

Alkyl halide $CH_3CH_2O^{\ominus}$ $Na^{\oplus}$ Ether

(d)

$$CH_3I \xrightarrow{(CH_3)_3N} (CH_3)_4N^{\oplus}I^{\ominus}$$

Alkyl halide   Quaternary ammonium salt

(e)

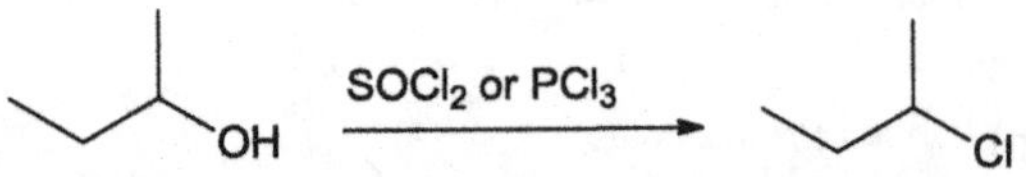

Alcohol $SOCl_2$ or PCl_3 Alkyl halide

Q 7.3. The rate dependency of an S_N2 reaction is on both the concentration of substrate and the nucleophile.

Rate = k (1-bromobutane)(hydroxide)

(a) If NaOH is doubled, so is the rate.
(b) If both NaOH and 1-bromobutane are doubled, the rate is quadrupled.
(c) If the volume is doubled, both NaOH and 1-bromobutane are halved. The rate is therefore lowered by a factor of 4.

Q 7.4.

(a)

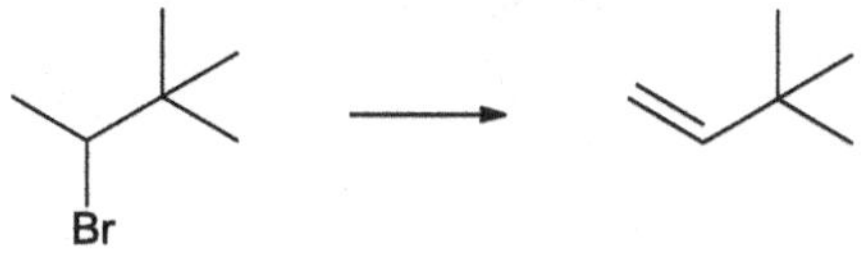
Br

(b)

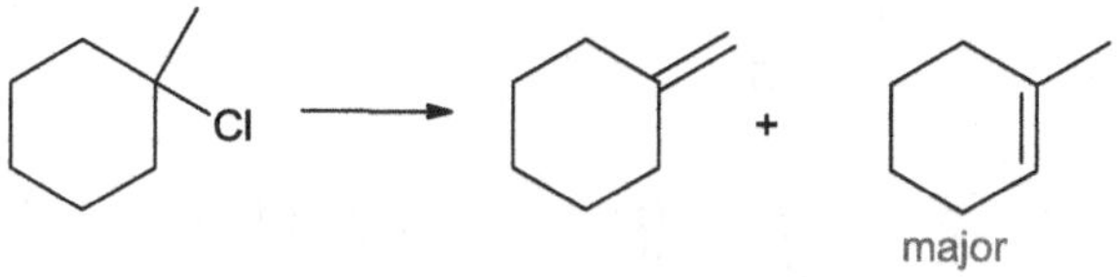
Cl
+
major

(c)

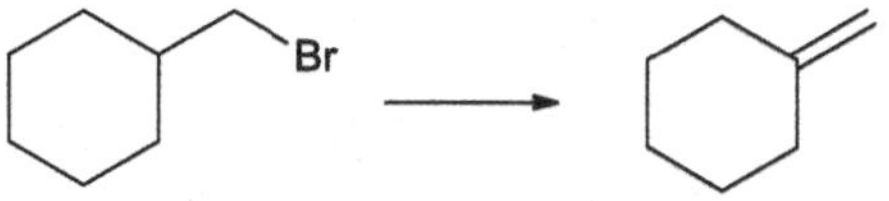
Br

(d)

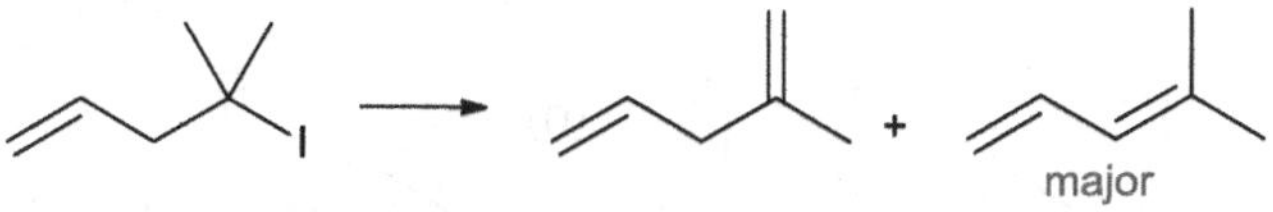
I
+
major

Q 7.5.

(a)

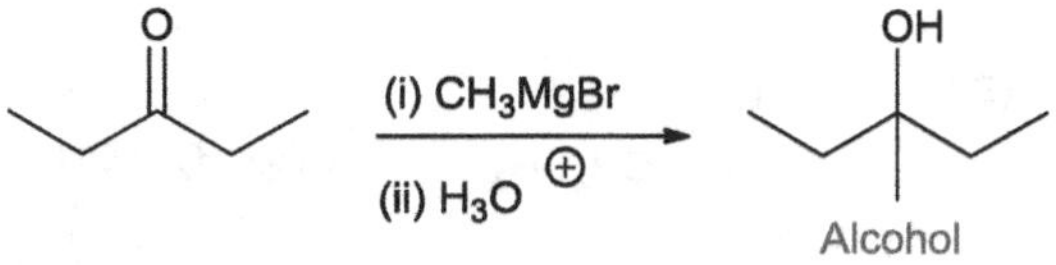
O
(i) CH3MgBr
(ii) H3O ⊕
OH
Alcohol

(b)

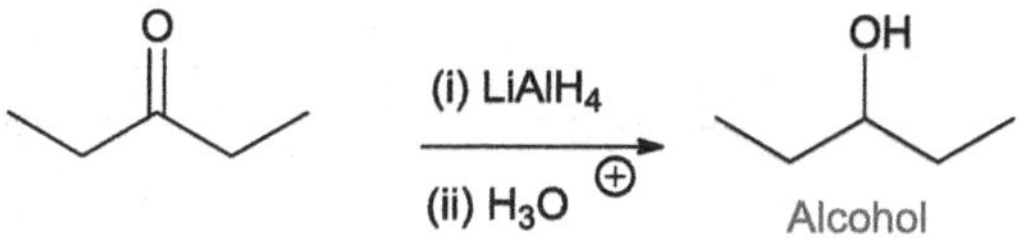
O
(i) LiAlH4
(ii) H3O ⊕
OH
Alcohol

(c)

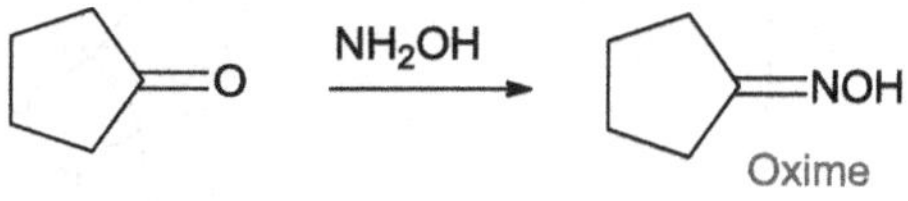
O
NH2OH
NOH
Oxime

(d)

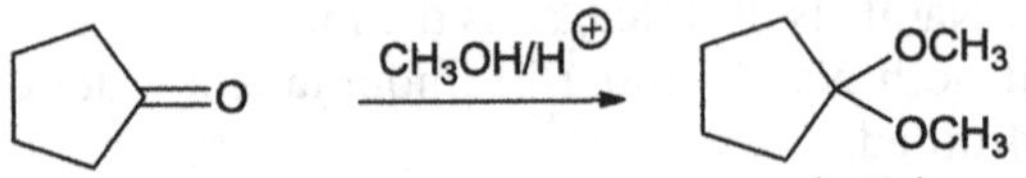

CH₃OH/H⊕

Acetal

(e)

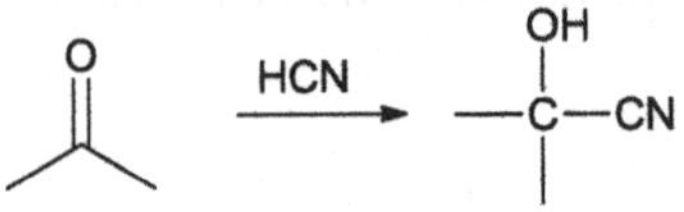

HCN

$$\underset{\text{Cyanohydrin}}{\overset{\overset{\text{OH}}{|}}{-\overset{|}{C}-CN}}$$

Q 7.6.

(a)

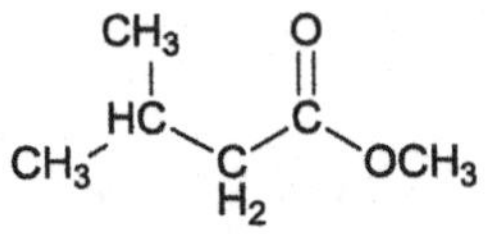

(b)

CH₃(CH₂)₆CN

(c)

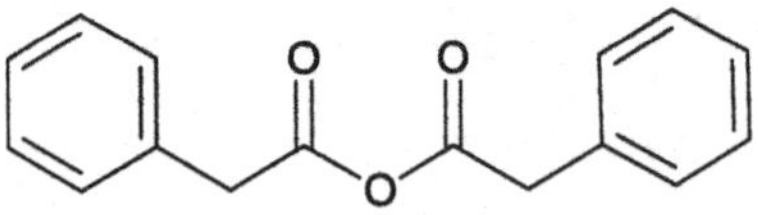

(d)

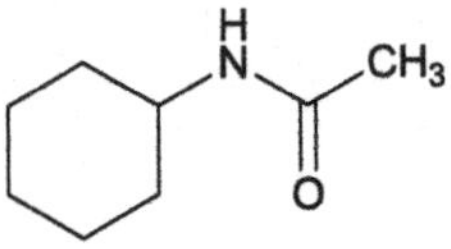

Q 7.7.

(a)

Carboxylic acid

(b)

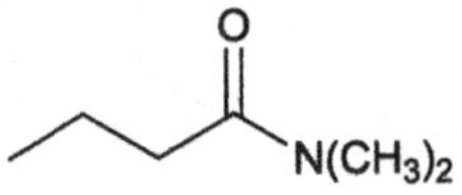

Amide

(c)

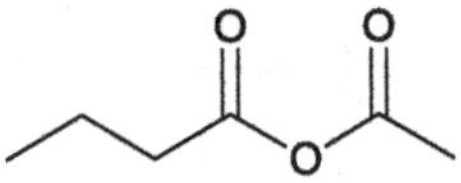

Acid anhydride

(d)

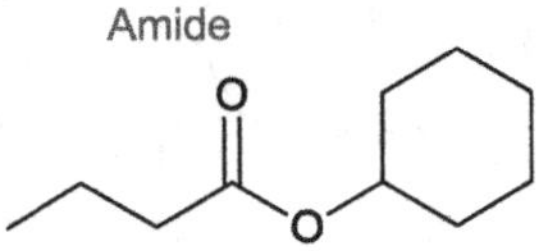

Ester

Q 7.8.

(a)

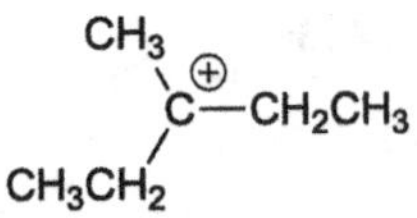

3° favoured

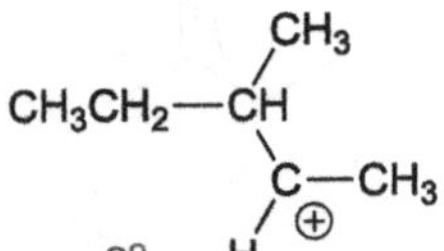

2°

(b)

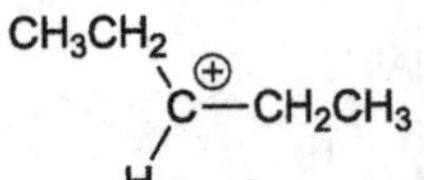

2°

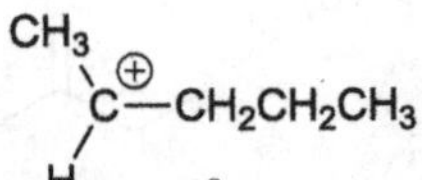

2°

(c)

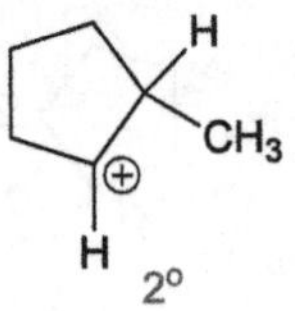

2°

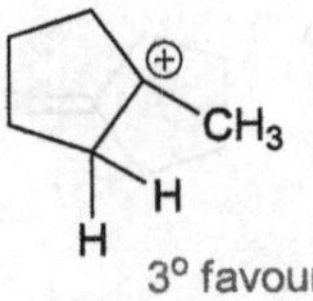

3° favoured

(d)

3° favoured

1°

Q 7.9.

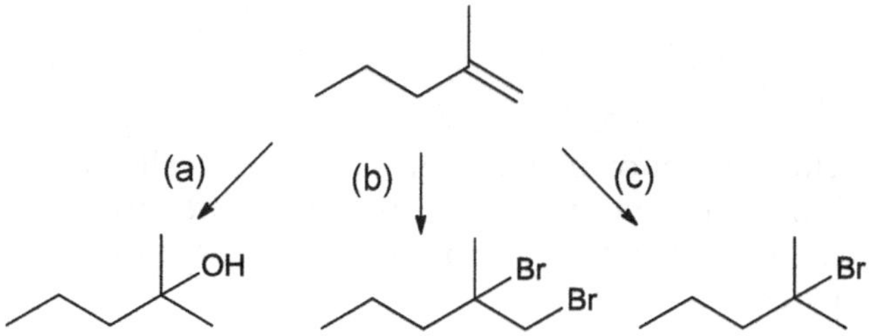

(a)

(b)

(c)

Q 7.10.

(a)/(b) $CH_3CH_2CH_2OH$ (c) $CH_3CH_2CO_2H$ (d) 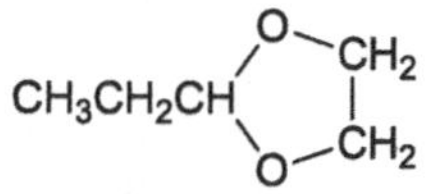 $CH_3CH_2CH\begin{smallmatrix}O-CH_2\\O-CH_2\end{smallmatrix}$

(e) $CH_3CH_2CH=NCH_2CH_3$ (f) $CH_3CH_2CO_2^{\ominus}$ + Ag°

Q 7.11.

(a) Cl_2/hv (b) Base (c) Br_2 (d) Base (e) HCl
        (NaOEt)/   (NaOEt)/heat
        heat

(f) HCl (g) $NaNH_2$ (h) Br_2 (1 (i) dil. (j) H_2/Lindlar
        followed by   equivalent) H_2SO_4/ catalyst (Pd/
        CH_3I $HgSO_4$ $BaCO_3$)

Q 7.12.

(a) $(CH_3)_2CHCH_2CHO$ $\xrightarrow{\text{LiAlH}_4}$ $(CH_3)_2CHCH_2CH_2OH$

    Aldehyde   Nucleophilic addition   Alcohol
                        (reduction)

(b)

$(CH_3)_2CHCH_2Br$ $\xrightarrow[\text{dilute}]{\overset{\oplus\ominus}{Na}\ OCH_2CH_3}$ $(CH_3)_2CHCH_2OCH_2CH_3$

Alkyl halide   Ether

            Nucleophilic substitution

(c) CH_3CH_2OH $\xrightarrow{\text{PCl}_3}$ CH_3CH_2Cl $\xrightarrow{CH_3OH}$ $CH_3CH_2OCH_3$

    Alcohol   Alkyl   Ether
                                halide
                Nucleophilic   Nucleophilic
                substitution   substitution

(d)

$$CH_3CH_2CHO \xrightarrow[\substack{H_2SO_4 \\ \text{Nucleophilic} \\ \text{addition}}]{NaCN} CH_3CH_2CH\overset{\displaystyle CN}{\underset{\displaystyle OH}{<}}$$

Aldehyde   Cyanohydrin

$$\xrightarrow[\substack{\text{Nucleophilic addition} + \\ \text{substitution (hydrolysis)}}]{\substack{\text{heat} \\ H_3O^{\oplus}}} CH_3CH_2CH\overset{\displaystyle CO_2H}{\underset{\displaystyle OH}{<}}$$

α-Hydroxy acid

Q 7.13.

(a) The difference between an alkene and alkane is the double bond. Electrophilic addition of bromine gives a color test. As the bromine is added to the alkene, the brown bromine color fades.

$$CH_3CH=C(CH_3)_2 \xrightarrow{Br_2} CH_3CH-C(CH_3)_2$$
$$\underset{\displaystyle Br \quad\; Br}{|\quad\;\; |}$$

(b) The difference between the two alcohols is the position of the hydroxyl group (regioisomers). Oxidation of the 2-OH isomer will give a methyl ketone that will undergo the haloform (usually iodoform) reaction. The standard conditions of the iodoform reaction will carry out the oxidation (I_2 will act as oxidant). The yellow iodoform precipitate is seen.

$$CH_3CH_2CH_2CHCH_3 \xrightarrow[NaOH]{I_2 \text{ (excess)}} \left[CH_3CH_2CH_2CCH_3 \right]$$
$$\underset{\displaystyle OH}{|} \qquad\qquad\qquad\qquad \underset{\displaystyle O}{||}$$

Methyl ketone

$$CH_3CH_2CH_2CO_2^{\ominus} + CHI_3\downarrow$$

Carboxylate   Iodoform

(c) Two approaches are possible. The first uses the acid–base reaction of the carboxylic acid with bicarbonate or carbonate to give bubbles of CO_2.

$$(CH_3)_2CHCO_2H \xrightarrow{NaHCO_3} (CH_3)_2CHCO_2^{\ominus} + CO_2\uparrow$$

Secondly, the aldehyde carbonyl group can be identified by the characteristic condensation (addition + elimination) reaction of 2,4-dinitrophenylhydrazine (Brady's reagent). This gives a precipitate of yellow/orange hydrazone.

2,4-dinitrophenylhydrazine

2,4-dinitrophenylhydrazone

The aldehyde can also be oxidized to the carboxylic acid. If this is carried out with ammonia silver complex $[Ag(NH_3)_2^+]$, a silver mirror deposit of metallic silver is seen as the silver is reduced.

$$(CH_3)_2CHCHO \xrightarrow{\overset{\oplus}{Ag(NH_3)_2}} (CH_3)_2CHCO_2^{\ominus} + Ag^{\circ} \downarrow$$

Carboxylate

(d) The difference between amines and amides involves the reactivity of the nitrogen lone pair. Amines are relatively basic while the amide lone pair is delocalized into the adjacent carbonyl group. The diazotization reaction with nitrous acid (HNO_2) is only effective with the basic amine. This case involves a primary amine and the product is an alcohol with the visual observation of bubbles of N_2 gas.

$$CH_3CH_2CH_2NH_2 \xrightarrow{HNO_2} CH_3CH_2CH_2OH + N_2 \uparrow$$

Alcohol

(e) This problem can be solved either from the alcohol or the ketone. Alcohols react readily with metallic sodium to release bubbles of H_2 gas.

Alcoholate

Alternatively the ketone would be detected by the yellow/orange hydrazone formed with Brady's' reagent as in (c) above.

(f) Again, the difference can be seen with either substrate. The alcohol would respond to metallic sodium as in (e) above and H_2 would bubble off. Alternatively, the alkyl halide would give a white precipitate of silver bromide on treatment with silver nitrate.

Q 7.14.

    (a) [D]   (b) [J]   (c) [L]
    (d) [I]   (e) [F]   (f) [G]

Q 7.15.

    (a) [I]   (b) [K]   (c) [E]
    (d) [D] + [J]   (e) [F]   (f) [L]

Q 7.16.

    (a) [K] + [F]   (b) [B]   (c) [M]   (d) [J] + [D]

Q 7.17.

Q 7.18.
    (a) Free radical substitution.
    (b) Yes, there is a chiral carbon present in the product.
    (c)

Q 7.19. The electron rich nature of the C=C makes attack by a nucleophile unfavorable. This is especially true in this case where the C=C has two

+I effects from the CH_3 groups. Electrophilic addition would be the reaction of choice.

Q 7.20.

(a)

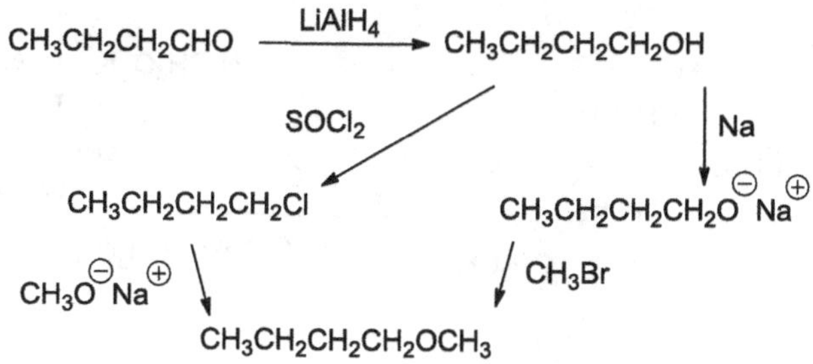

$$CH_3CH_2CH_2CHO \xrightarrow{\text{LiAlH}_4} CH_3CH_2CH_2CH_2OH$$

SOCl$_2$

Na

$$CH_3CH_2CH_2CH_2Cl \qquad CH_3CH_2CH_2CH_2O^{\ominus} \ Na^{\oplus}$$

$CH_3O^{\ominus} \ Na^{\oplus}$

CH_3Br

$$CH_3CH_2CH_2CH_2OCH_3$$

(b)

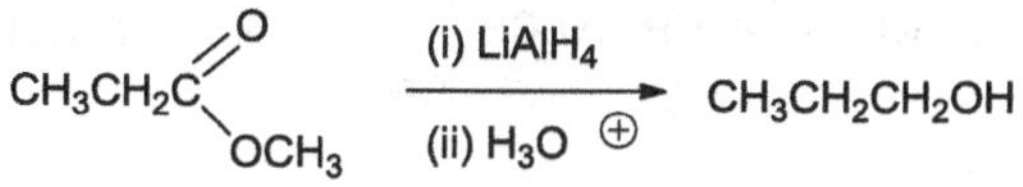

$$CH_3CH_2C\!\!\begin{array}{c}{}^{\displaystyle O}\\[-2pt]{}\end{array}\!\!\!\diagdown_{OCH_3} \xrightarrow[\text{(ii) } H_3O^{\oplus}]{\text{(i) LiAlH}_4} CH_3CH_2CH_2OH$$

(c)

$$CH_3CH_2OCH_3 \xrightarrow[\text{heat}]{\text{conc HI}} CH_3CH_2I \xrightarrow[\text{heat}]{CH_3CH_2O^{\ominus} K^{\oplus}} CH_2\!\!=\!\!CH_2$$

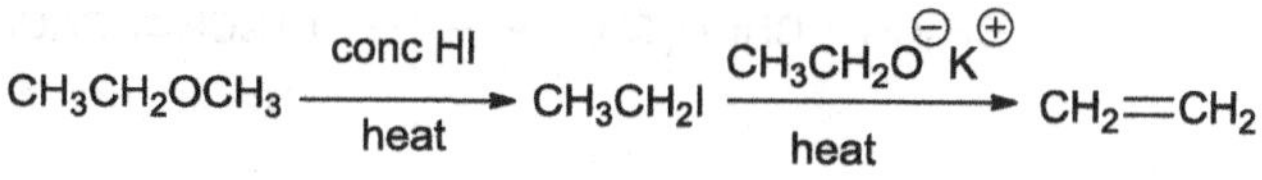

(d)

$$CH_3CH_2CH_2\overset{\displaystyle O}{\overset{\|}{C}}CH_3 \xrightarrow[\text{(ii) } H_3O^{\oplus}]{\text{(i) NaOH/I}_2} CH_3CH_2CH_2CO_2H$$

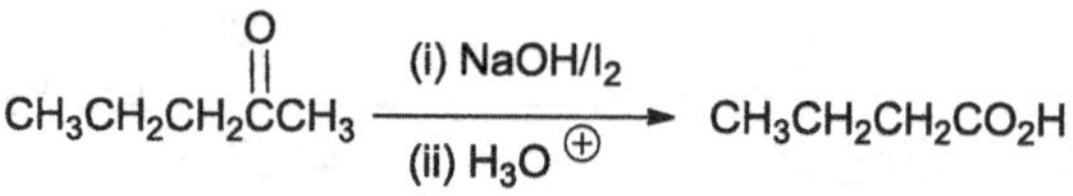

Q 7.21.

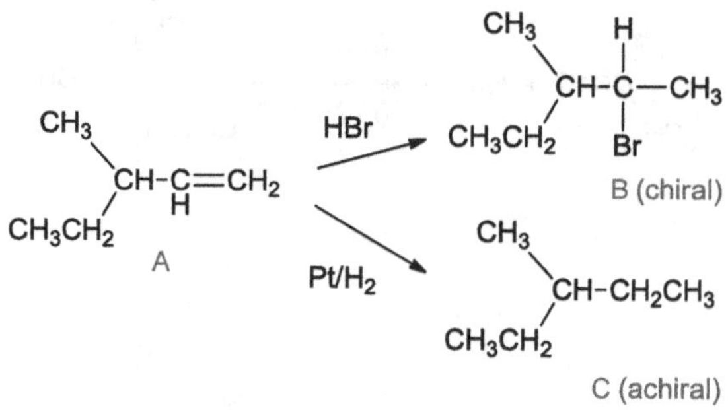

Q 7.22.

$$(CH_3)_2CHOCH(CH_3)_2 \xrightarrow{\text{HI/heat}} (CH_3)_2CHI + H_2O$$

NaCN

$$(CH_3)_2CHCO_2H \xleftarrow[\text{heat}]{H_3O^{\oplus}} (CH_3)_2CHCN + NaI$$

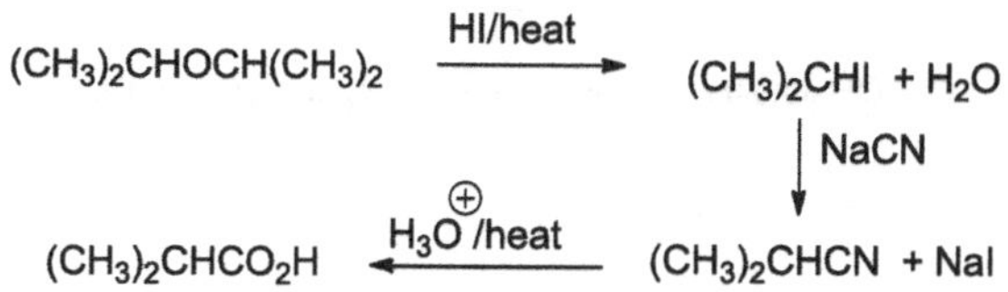

Q 7.23.

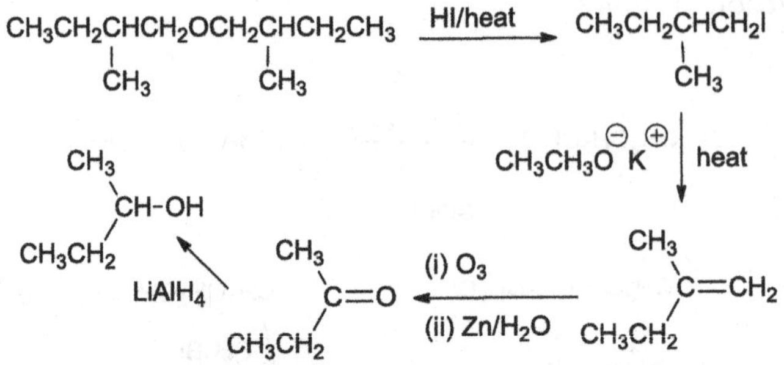

$$CH_3CH_2CHCH_2OCH_2CHCH_2CH_3 \xrightarrow{HI/heat} CH_3CH_2CHCH_2I$$

with CH$_3$ substituents on left structure (two CH$_3$ groups) and CH$_3$ on right product.

$$CH_3CH_2O^{\ominus} K^{\oplus} \Big| heat$$

CH$_3$
CH–OH
CH$_3$CH$_2$ $\xleftarrow{LiAlH_4}$

CH$_3$
C=O
CH$_3$CH$_2$ $\xleftarrow[\text{(ii) Zn/H}_2\text{O}]{\text{(i) O}_3}$

CH$_3$
C=CH$_2$
CH$_3$CH$_2$

Q 7.24.

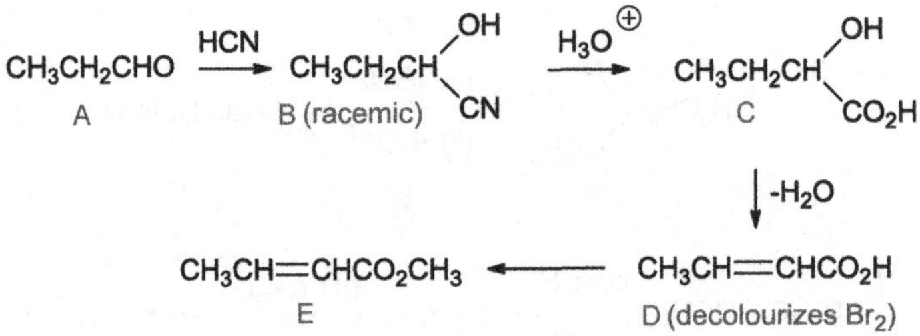

$$CH_3CH_2CHO \xrightarrow{HCN} CH_3CH_2CH\overset{OH}{\underset{CN}{|}} \xrightarrow{H_3O^{\oplus}} CH_3CH_2CH\overset{OH}{\underset{CO_2H}{|}}$$

A   B (racemic)   C

$$\Big| -H_2O$$

$$CH_3CH=CHCO_2CH_3 \longleftarrow CH_3CH=CHCO_2H$$
E   D (decolourizes Br$_2$)

Q 7.25.

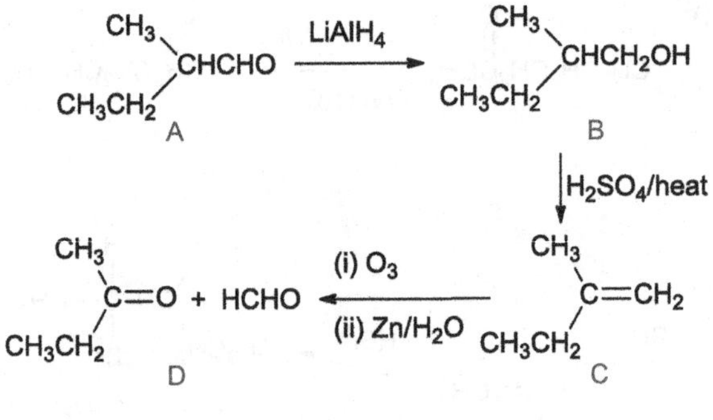

CH$_3$
CHCHO
CH$_3$CH$_2$
A $\xrightarrow{LiAlH_4}$
CH$_3$
CHCH$_2$OH
CH$_3$CH$_2$
B

$$\Big| H_2SO_4/heat$$

CH$_3$
C=O + HCHO
CH$_3$CH$_2$
D $\xleftarrow[\text{(ii) Zn/H}_2\text{O}]{\text{(i) O}_3}$
CH$_3$
C=CH$_2$
CH$_3$CH$_2$
C

Q 7.26.

CH$_3$
C=CH$_2$
CH$_3$CH$_2$
A $\xrightarrow[\text{(ii) Zn/H}_2\text{O}]{\text{(i) O}_3}$
CH$_3$
C=O + HCHO
CH$_3$CH$_2$
B

$$\Big| NaOH/I_2$$

$$CH_3CH_2CO_2H \xleftarrow{NaHCO_3} CH_3CH_2CO_2^{\ominus} Na^{\oplus}$$
C

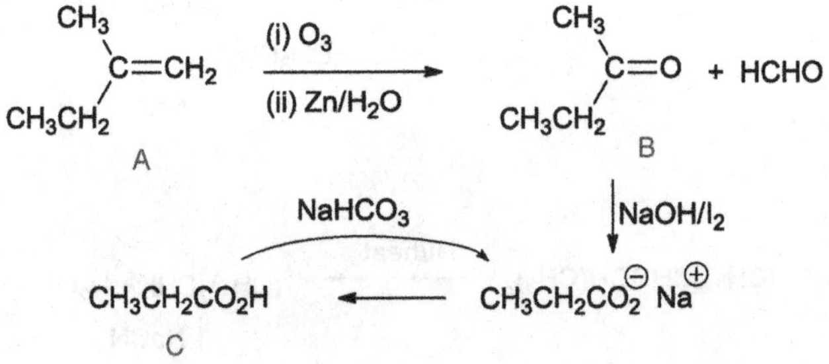

Q 7.27.

Q 7.28.

CHAPTER 8: NATURAL PRODUCT BIOMOLECULES

Q 8.1.

Ketopentose   Aldopentose   Aldoheptose

Q 8.2.

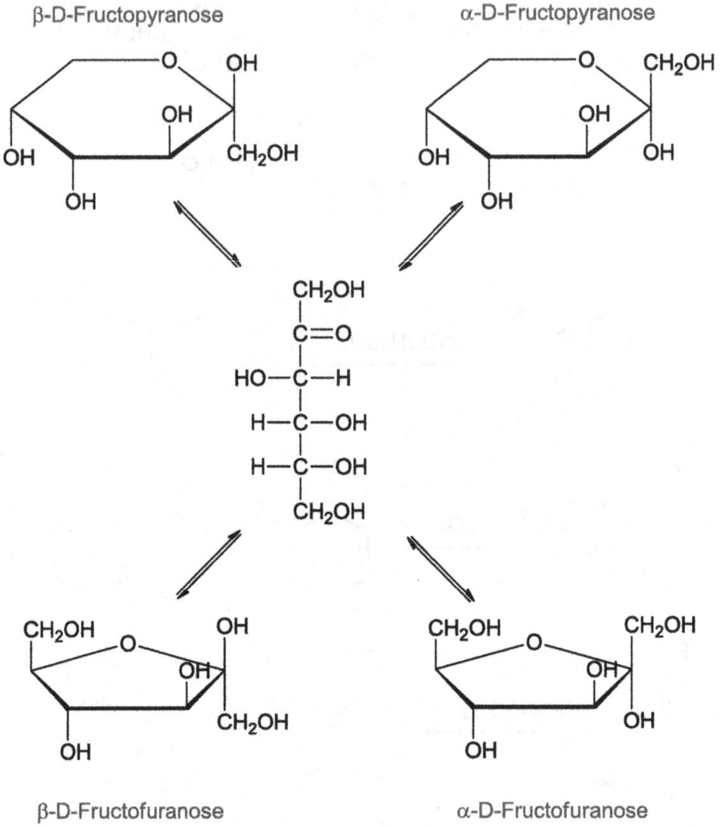

β-D-Fructopyranose

α-D-Fructopyranose

CH₂OH
C=O
HO—C—H
H—C—OH
H—C—OH
CH₂OH

β-D-Fructofuranose

α-D-Fructofuranose

Q 8.3. The two monosaccharides that contain oxidizable aldehyde groups (the aldoses) will act as reducing agents with copper (II) (Fehling's or Benedict's solutions) or silver (I) (Tollen's solution) ions. The remaining example is a 2-ketose that will also undergo oxidation after enol tautomerization to the corresponding aldose.

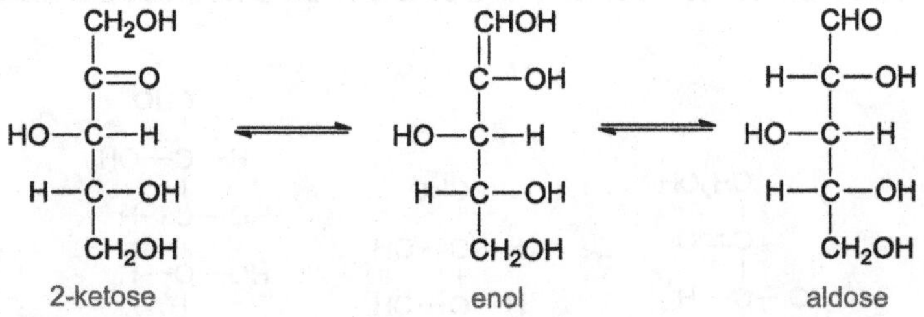

CH₂OH
C=O
HO—C—H
H—C—OH
CH₂OH

2-ketose

CHOH
C—OH
HO—C—H
H—C—OH
CH₂OH

enol

CHO
H—C—OH
HO—C—H
H—C—OH
CH₂OH

aldose

Q 8.4. A total of six isomers are possible.

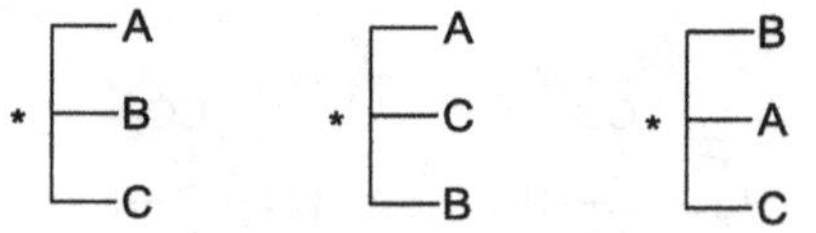

where A = $CH_3(CH_2)_{14}CO_2$ palmitic

B = $CH_3(CH_2)_{16}CO_2$ stearic

C = cis-$CH_3(CH_2)_7CH=CH(CH_2)_7CO_2$ oleic

Three regioisomers each existing as a pair of enantiomers at the chirality centre

Q 8.5. The difference is caused by the presence of the cis double bond that prevents the close approach of the hydrocarbon chains and therefore weakens the London dispersion force interactions.

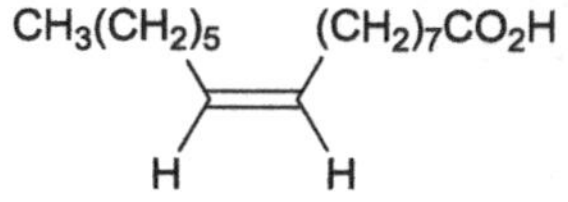

palmitoleic acid

Q 8.6.

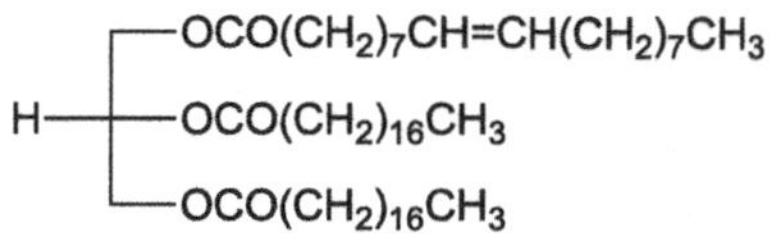

or the enantiomer

Q 8.7. A total of 6 moles of hydrogen are required. One for the oleic, two for the linoleic, and three for the linolenic.

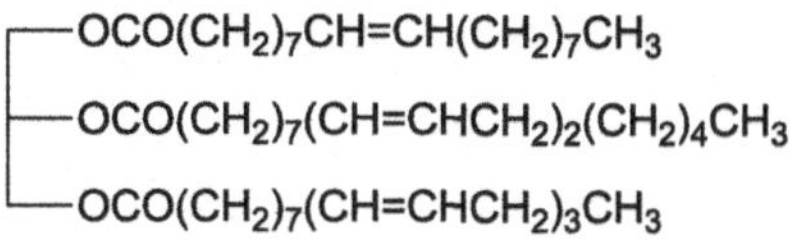

oleic

linoleic

linolenic

Q 8.8. They are generally soluble in nonpolar solvents because of the high hydrocarbon content.

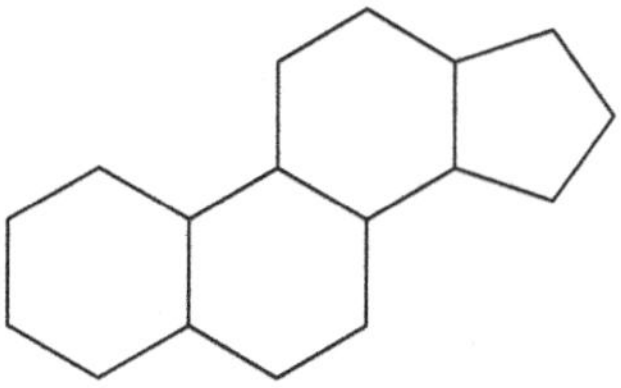

Q 8.9.

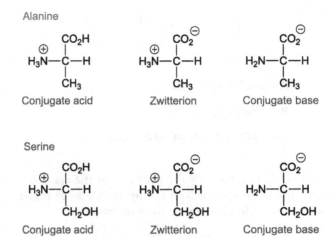

Alanine

Conjugate acid — Zwitterion — Conjugate base

Serine

Conjugate acid — Zwitterion — Conjugate base

Glossary of Technical Definitions

A,B

Achiral   has no chirality center
Activation energy   the energy needed for a reaction to proceed
Acyclic   structure that does not have a ring in it
Acyl group   a fragment attached at the carbon of a carbonyl group
Acylation   the addition of an acyl group
Addition   a reaction type that adds substituents to multiple bonds
Aliphatic   any hydrocarbon that is not aromatic
Alkyl group   derived from an alkane by removing one hydrogen substituent
Alkylation   the addition of an alkyl group
Aromatic   a class of compound that has a special delocalized electron bonding system
Asymmetric   has no symmetry
Bimolecular   involving two different species (molecules)
Branched chain   a carbon framework that has sidechains
Brønsted–Lowry   defines an acid as a proton donor, and a base as a proton acceptor

C

Carbanion   a fragment where carbon carries a negative charge
Carbocation   a fragment where carbon carries a positive charge
Carbocyclic   a ring that is made of only carbon atoms
Carbonyl group   a carbon with a double bond to oxygen
Chirality   the ability of a molecule to give non-superimposable mirror images
Chirality (chiral) center   a tetrahedral center with four different substituents
Combustion   the burning of a substance in the presence of oxygen
Condensation (reaction)   an overall reaction that combines an addition with an elimination
Configuration (absolute configuration)   the exact three-dimensional arrangement about a tetrahedral center
Conformational isomers (conformers)   differ simply by rotation around a single bond
Conjugated (system)   alternating single and multiple bonds
Concerted   a single step process
Coplanar   lying in the same plane of geometry
Cumulative (system)   multiple bonds joined directly to each other
Curly arrows   a convenient way to show electron movement
Cyclic compound   a structure that has a ring in it

D

Dehydration   the loss of the components of water
Delocalized   refers to bonding electrons that do not have a well-defined fixed position
Deprotonation   the removal of a proton by a base
Derivative (specific)   made from a substance by reaction with a standard reagent
Double bond equivalents   multiple bonds or rings that decrease the number of substituents needed to complete tetravalency in organic compounds

E,F

Eclipsed   a conformation where substituents on bonded centers rotate to line up with each other
Electrophile   an electron-poor species
Electronegativity   measures the attraction a bonded atom has for the bonding electrons
Elimination   a reaction type where the loss of substituents forms a multiple bond
Enantiomers (optical isomers)   a pair of non-superimposable mirror images
Endothermic process   one that consumes energy
Energy diagram   qualitative picture of energy changes during a reaction
Enol   an equilibrium form of a carbonyl compound

Enolate   the deprotonated form of an enol
Epoxidation   the addition of a bridging oxygen atom to a multiple bond
Exothermic process   one that gives off energy
Formula   the type and number of each atom that is found in a molecule
Functional group   a fixed arrangement of atoms that defines a compound's properties
Functional class   structures that all have the same functional group

G,H

Geometric (*cis–trans*) isomers   different geometries caused by a C=C or a ring restricting bond rotation
Heteroatom   atoms other than carbon and hydrogen
Heterocyclic   a ring that has at least one noncarbon atom
Heterogenic bond making   when cations bond with anions
Heterolysis   bond breaking that results in a cation and an anion
Homologous series   set of compounds formed by adding a $-CH_2-$ group to each preceding member
Homolysis   bond breaking that results in two radicals
Homogenic bond formation   when radicals bond
Hybridization   a simple model to explain single and multiple chemical bonding and molecular shape
Hybrid atomic orbital   the result of mixing atomic orbitals by hybridization
Hydration   the addition of the elements of water (across a multiple bond)
Hydrocarbon   compound composed of only carbon and hydrogen
Hydrogenation   the addition of hydrogen to reduce a multiple bond
Hydrolysis   a decomposition reaction of a substrate by water
Hydrophilic   soluble in water
Hydrophobic   insoluble in water
Hydroxylation   the addition of two hydroxyl groups to a multiple bond

I–L

Inductive effect (I)   shows the ability and direction of polarization of the electrons in a sigma bond
Initiation   the formation of radicals needed to carry out a radical process
Intermolecular   between separate molecules
Intramolecular   within the same molecule
Isolated (system)   multiple bonds that are more than one single bond apart
Isomers   having the same molecular formula, but with the atoms differently arranged
IUPAC   International Union of Pure and Applied Chemistry
Lewis   defines an acid as an electron pair acceptor, and a base as an electron pair donor
Localized   refers to bonding electrons found in a well-defined fixed space

M,N

Mirror image   the reflection of a molecule in a mirror
Molecule   a group of bonded atoms
Multiple bond   where two bonding atoms share more than one electron pair for bonding
Natural product   compound that is made in nature by a living organism
Nucleophile   an electron-rich species

O–Q

Octet   the number of electrons needed to have a full valence shell
Organometallic   a substance that has both an organic and a metallic part
Oxidation number (state)   number of electrons gained or lost by an element when bonded in a compound
Ozonolysis   the oxidative breaking of a multiple bond
Pi (π) bond   when bonding orbitals overlap at right angles to the orbital axes

Plane polarized light   light that has been filtered to allow only waves in a single plane
Polar bond (polarization)   where one nucleus attracts the bonding electrons more than the other
Polarimeter   instrument to measure the effect of molecular structure on plane polarized light
Propagation   the repeated carrying out of a radical process
Protonation   the addition of a proton
Qualitative   not an absolute number, used to give relative indication of a property

R

Racemic mixture   an equal mixture of a pair of enantiomers
Radical   a fragment that has an unpaired valence electron
Rate-determining step   the slowest step in a stepwise process
Reactive intermediate   a proven species that is involved in a reaction mechanism
Reaction mechanism   detailed description of how bonds are broken and made in a reaction
Reaction equilibrium   gives the overall direction of reaction
Reaction rate   how fast a reaction occurs
Rearrangement   a reaction type that leads a structure to reform as a structural isomer
REDOX reaction   a reaction that involves reduction and oxidation
Resonance   a concept to help explain bonding when a single Lewis diagram is not accurate
Resonance forms   alternative Lewis structures for a molecule
Resonance hybrid   a weighted average of possible resonance forms

S

Saturated   has no multiple bonds
Sequence rules   a series of rules that must be used in a specific order
Sigma (σ) bond   when bonding orbitals overlap along the orbital axes
Single (covalent) bond   where two atoms share an electron pair for bonding
Solvation   interaction between solvent and dissolved molecules/ions
Staggered   a conformation where substituents on bonded centers rotate as far apart as possible
Stepwise (reaction)   involves more than one step, one after the other
Stereochemistry   the three-dimensional nature of molecules
Stereoisomer   molecules with the same order of atomic bonding, but different arrangements in space
Straight chain   a carbon framework that has no branching
Structural isomers   where the atoms are joined together in different orders
Structure   the arrangement in space of the bonded atoms in a molecule
Substituents   term for an atom or group of atoms attached by a bond to an atom of interest
Substitution   a reaction type where one substituent is replaced by another
Superimposed   placed on top of to match exactly

T–V

Tautomerism   equilibrium of structural isomers differing only in the position of an acidic hydrogen
Terminal   at the end of a chain
Termination   the trapping of radical to end a radical process
Tetravalent   needs to form four bonds to complete a valence shell octet
Three-dimensional (3D) shape   as given by the plane of paper and the directions in or out of the paper
Trace element   an element needed in very small amounts in biological systems
Transition state   a proposed species involved in a reaction mechanism
Unimolecular   involving only a single species (molecule)
Unsaturated   has at least one multiple bond
Valence (shell)   the outer shell of an atom

Index

219

Printed in the United States
By Bookmasters